Daily Warm-Ups
ALGEBRA
NCTM Standards

Betsy Berry, Ph.D.
Indiana University Purdue University Fort Wayne

The classroom teacher may reproduce materials in this book for classroom use only.
The reproduction of any part for an entire school or school system is strictly prohibited.
No part of this publication may be transmitted, stored, or recorded in any form
without written permission from the publisher.

1 2 3 4 5 6 7 8 9 10
ISBN 978-0-8251-6453-8
Copyright © 2008
J. Weston Walch, Publisher
40 Walch Drive • Portland, ME 04103
www.walch.com
Printed in the United States of America

Table of Contents

Introduction v

About the CD-ROM vi

Part 1: Developing Algebraic Thinking Through and for
      Mathematical Modeling 1

Part 2: Analyzing Change in Various Contexts 53

Part 3: Using Algebraic Symbols I 68

Part 4: Using Algebraic Symbols II 93

Part 5: Understanding Patterns, Relations, and Functions 127

    Answer Key 181

Daily Warm-Ups: Algebra, NCTM Standards

About the CD-ROM

Daily Warm-Ups: Algebra, NCTM Standards is provided in two convenient formats: an easy-to-use, reproducible book and a ready-to-print PDF on a companion CD-ROM. You can photocopy or print activities as needed, or project them on a large screen via your computer.

The depth and breadth of the collection gives you the opportunity to pick and choose specific skills and concepts that correspond to your curriculum and instruction. The activities address all of the NCTM Standards for Algebra, including Algebra I, Algebra II, and some trigonometry. Use the table of contents and the information on the part title pages to help you select appropriate tasks.

Suggestions for use:

- Choose an activity to project or print out and assign.
- Select a series of activities. Print the selection to create practice packets for learners who need help with specific skills or concepts.

Daily Warm-Ups: Algebra, NCTM Standards

Introduction

I taught mathematics for many years at the middle- and high-school level. I know how precious every moment can be in a class period. The problems, exercises, activities, reflections, and writing prompts in this collection of warm-ups are just what I wish I had during those years. These warm-ups are organized in five parts around four strands of algebra designated by the National Council of Teachers of Mathematics *Principles and Standards for School Mathematics*: using mathematical models; analyzing change; using algebraic symbols; and understanding patterns, relations, and functions. The NCTM process standards, especially communication and representation, are also integrated within these activities.

The warm-ups address all algebra standards within NCTM, and exceed Algebra I, include Algebra II, and touch upon Trigonometry. The NCTM expectations within each topic area for grades 6 through 8 and grades 9 through 12 are addressed at least twice in different ways. Some problems address more than one standard or expectation, but have been placed in the respective section according to the topic area that I thought was best addressed by the activity.

The warm-ups are organized by standards rather than by level of difficulty. Use your judgement to select appropriate problems for your students. The problems are not meant to be done in consecutive order from the beginning of the book to the end. Some of these problems are stand-alone, some can launch a topic, some can be used for journal prompts, and some refresh students' skills and concepts. All are meant to enhance and complement high-school algebra curriculum programs. They do so by providing resources for teachers for those short spaces of time of 5 to 15 minutes when class time might go unused.

—Betsy Berry, Ph.D.

Daily Warm-Ups: Algebra, NCTM Standards

Part 1: Developing Algebraic Thinking Through and for Mathematical Modeling

National Council of Teachers of Mathematics: "Instructional programs from pre-kindergarten through grade 12 should enable all students to use mathematical models to represent and understand quantitative relationships."

Expectations

Grades 6–8:

- Model and solve contextualized problems using various representations, such as graphs, tables, and equations.

Grades 9–12:

- Identify essential quantitative relationships in a situation and determine the class or classes of functions that might model the relationships.
- Use symbolic expressions, including iterative and recursive forms, to represent relationships arising from various contexts.
- Draw reasonable conclusions about a situation being modeled.

Patricia's Patterns

Patricia has been exploring number patterns consisting of 4 consecutive integers. She has noticed a relationship between the products of the inside numbers and the outside numbers. Complete Patricia's table and then tell what you think Patricia's discovery is. Will this always be true? Justify your answer algebraically.

Number	Inside product	Outside product
3, 4, 5, 6		
6, 7, 8, 9		
10, 11, 12, 13		
15, 16, 17, 18		
40, 41, 42, 43		

Daily Warm-Ups: Algebra, NCTM Standards

Stacking Rods 1

Cuisenaire rods are wooden materials often used for mathematics activities in elementary classrooms. They were invented by Georges Cuisenaire in 1952. They consist of 10 rectangular rods with height = 1 cm, width = 1 cm, and lengths of 1–10 cm. Most Cuisenaire rods follow the system below:

white rod = 1 cm
red rod = 2 cm
light green rod = 3 cm
purple rod = 4 cm
yellow rod = 5 cm
dark green rod = 6 cm
black rod = 7 cm
brown rod = 8 cm
blue rod = 9 cm
orange rod = 10 cm

Pierre is making stacks of purple rods and recording the surface area of the exposed sides in a table. What is the surface area of a stack of n purple rods? (Assume that the sides of the rods are exposed, but not the bottom surface.)

Daily Warm-Ups: Algebra, NCTM Standards

Stacking Rods II

Marie is investigating purple Cuisenaire rods that are stacked in a staircase.

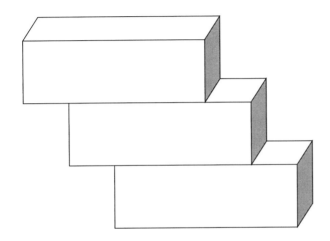

The rods are displaced in each step by exactly 1 unit. Remember that the purple rod is 4 units long. Find the surface areas of stacks of different numbers of rods. Then find a rule or a formula for finding the surface area of a staircase of n number of rods. (Assume that the bottom face of each rod is unexposed.)

Daily Warm-Ups: Algebra, NCTM Standards

Areas of Circles and Squares

Plot a scatter plot that shows the relationship between the length of one side of a square and the area of the square. Plot another scatter plot (possibly on the same set of axes) that shows the relationship between the radius of a circle and the area of the circle. How would you describe these scatter plots? How are they similar? How are they different? Show tables of values and your scatter plots as part of your solution.

Water Balloon Bungee Jump

Natalie and Charlotte are conducting an experiment using rubber bands and water balloons. Their challenge is to predict the distance a water balloon will fall for any number of rubber bands that they might use. This is the data that they have collected so far:

Number of rubber bands	3	4	5	6	7	8	9	10	11
Distance fallen (cm)	71	83	95	115	128	144	160	175	190

They plan to test their conclusions from this data by dropping a water balloon from a stairwell at a starting height of 12.5 feet. How many rubber bands should they use to get the water balloon as close to the floor as possible, but not break it?

Daily Warm-Ups: Algebra, NCTM Standards

How Many Handshakes?

It is a tradition in the Magnolia Magnet School of Meticulous Manners for all students and teachers in a classroom to shake hands with every other person in the room. On a typical Monday morning, there are 16 people in Mr. Manners' classroom. How many handshakes will there be? Find a way to determine how many handshakes there will be in any classroom on any day.

Daily Warm-Ups: Algebra, NCTM Standards

Spring Break Flight Costs

Ray Airhart is chief pilot for Eagle Valley Aviation. He is promoting a charter flight to Florida for the Eagle Valley High School senior class spring break. The flight will cost $200 per person for 10 students or fewer. For every 4 additional people who go on the trip, he can lower the price for each person an additional $16. How many people will make the most money for Eagle Valley Aviation?

Daily Warm-Ups: Algebra, NCTM Standards

Pet Mice

Linn has been keeping several mice as pets for some time. She has recently learned that she can raise mice to sell to pet stores. However, she would need to provide the mice in lots of at least 200. To predict when she will have enough mice to sell, she has collected the data in the chart below.

Month	Mice
1	3
2	7
3	17
4	36
5	65

1. Make a scatter plot of Linn's data and draw a line of best fit.
2. Find an equation in slope-intercept form for your line.
3. Is Linn correct in using a linear model to predict when she will have 200 mice? If so, predict when. If not, explain why not.

Daily Warm-Ups: Algebra, NCTM Standards

Relating Temperature and Altitude

Keri has just begun pilot lessons. She is very interested in the relationships among altitude, pressure, temperature, and density. She did some online research and found the table below. It gives the standard values of pressure, temperature, and density at different altitudes.

Altitude (ft)	Pressure (Hg)	Temperature (°F)	Density (%)
sea level	29.92	59.0	100
2,000	27.82	51.9	94.3
4,000	25.84	44.7	88.8
6,000	23.98	37.6	83.6
8,000	22.22	30.5	78.6
10,000	20.57	23.3	73.8
12,000	19.02	16.2	69.3
14,000	17.57	9.1	65.0
16,000	16.21	1.9	60.9

Use a graphing calculator to make a graph of the relationship between altitude and temperature for $0 \leq A \leq 20{,}000$. Find a regression equation that you think works for this data. Be ready to explain your thinking.

Daily Warm-Ups: Algebra, NCTM Standards

© 2008 Walch Education

Fencing Fernando

Franny lives on a small farm with many unusual animals. She has just been given a Vietnamese potbellied pig named Fernando. But before she can get Fernando, she needs to build a pen for him. If needed, she can use one side of a long barn that measures 200 feet. She also knows that she has 120 feet of fencing available. Answer the questions below about Fernando's pen. Be prepared with drawings and mathematical models to justify your answers.

1. What are the longest sides that a rectangular pen could have if the barn is used for one side?
2. Suppose Franny wants a square pen. What are the dimensions of the square, if all the fencing and one side of the barn is used?
3. What is the maximum area that Franny can create for her pig?
4. What is the minimum area that Franny can create?
5. Suppose Franny decides to build her pen in a field, using all of the fencing but not the side of the barn. What are the maximum and minimum areas of her pen?

Daily Warm-Ups: Algebra, NCTM Standards

Boxes, Little Boxes

Jaden is exploring the characteristics of small boxes. He is creating boxes using one sheet of construction paper for each box. He makes the box by cutting out squares in 1-inch increments from the four corners of the paper and then folding up the edges. If the paper that Jaden is using is 9 inches by 12 inches, what are the sizes of the boxes that he is designing? Create a table of values that records the length, width, and volume of all the boxes you think Jaden could create in this way. Also include the length of the square cutouts, the surface area of the box, and any other data of interest to you. Do you see patterns in your data? What is the length of the square cutouts for the box with the greatest volume? With the smallest volume? If s represents the side of the square that Jaden cuts, how could he represent the volume of the resulting box in terms of s?

The Librarian's Dilemma

Margo is a librarian for the east branch of the Harrison City Library. Each year she adds new books, both fiction and nonfiction, to the library's collection. This year her budget limits her to no more than 75 new books. Library policy states that new fiction books can be no more than half the number of nonfiction books.

1. If x represents the number of nonfiction books and y represents the number of fiction books, write a system of inequalities that models Margo's situation.

2. Graph this system of inequalities.

3. What information is revealed in your graph?

4. What is the greatest number of fiction books that Margo can buy this year?

Daily Warm-Ups: Algebra, NCTM Standards

Pizza Toppings

You and a group of friends are planning to order two large pizzas and some soft drinks. Each pizza costs $12.00. Each extra topping costs $0.50, and each soft drink costs $1.00. You have a total of $40.00 to spend.

Represent this situation as a feasible region on a graph, and indicate the corner points of the region and its boundaries.

Daily Warm-Ups: Algebra, NCTM Standards

Working for Uncle Daniel

Your Uncle Daniel has proposed an unusual payment scheme for a summer job that you are planning to do for him. He says that he will pay you $1,000 in the morning on the first day, but at the end of the day you must pay him a commission of $100. The next day he will double what you cleared the first day, but you must double what you pay him for commission. This arrangement will go on for as long as you wish to work for him. Create a table, a graph, and an argument to present to your uncle that explains and justifies your decision.

Daily Warm-Ups: Algebra, NCTM Standards

The Growing Cube

A cube is having its first birthday. This cube is called a *unit cube*. Describe how many corners (vertices), faces, and edges it has. On the cube's second birthday, its edges are twice as long as on its first birthday. How many unit cubes would be needed to equal the volume of this new 2-year-old cube? On the cube's third birthday, its edges are 3 times as long as those of the unit cube. This pattern continues for each birthday. How many unit cubes would be needed to equal the volume of a cube that is 3, 4, 5, 6, 7, 8, 9, and 10 years old? Write a rule to calculate the number of unit cubes needed for any number of birthdays. Then draw a graph of the number of unit cubes as a function of the number of years old the cube is.

Coloring Cubes

Crystal and T.J. are considering cubes of many sizes that have been dipped completely in paint. They start by thinking about a cube that is 10 units long on each side. After it is dipped and dries, the 10-unit cube is cut into single-unit cubes.

1. How many of these newly cut unit cubes are painted on 3 faces, 2 faces, 1 face, and zero faces? Answer those same questions for cubes that are 2, 3, 4, 5, 6, 7, 8, and 9 units long on each side. Then make a chart of your findings for all the cubes. Look for patterns.

2. Where are the unit cubes with 3 painted faces located on the various cubes? With 2 painted faces? With 1 painted face? With zero painted faces?

3. Make a prediction for a cube that is 15 units long on one side. Describe the patterns using symbolic rules that could be used for any size cube.

Daily Warm-Ups: Algebra, NCTM Standards

Credit-Card Charges

Charles purchased a new flat-screen television for $2,000 with his credit card on July 1. His credit-card company charges 1.5% interest on the unpaid balance.

1. If Charles does not pay the credit-card company at the end of one month, how much will he owe?

2. How much will he owe at the end of two months?

3. How much will he owe at the end of three months?

4. When will his unpaid balance reach $2,500?

Daily Warm-Ups: Algebra, NCTM Standards

© 2008 Walch Education

Direct or Inverse?

Look at the data sets below. Decide if each one represents a situation that would give a direct proportion as a mathematical model, an inverse proportional representation, or neither. Explain your thinking for each one.

1.

x	3	−10	−2	2.5	20
y	1.67	−0.5	−2.5	0.2	0.25

2.

x	6	12	4	−3	24
y	2	1	3	−4	0.5

3.

x	−21	−3	6	12	15
y	−14	−2	4	8	10

4.

x	8	−12	−2	−3	4
y	$-\frac{1}{2}$	$\frac{-4}{3}$	−1	$\frac{-7}{3}$	0

Daily Warm-Ups: Algebra, NCTM Standards

Inverse Variation

1. Two quantities, x and y, vary inversely, and when $x = 4$, $y = 3$. Find the missing values for the points below that are also on the graph of this inverse variation.
   a. $(3, y)$
   b. $(x, 2)$
   c. $(1, y)$
   d. $(x, -4)$
   e. $(x, 24)$

2. Find five points that are on the graph of the equation $y = \dfrac{32}{x}$, and graph them and the equation to check your data points.

Daily Warm-Ups: Algebra, NCTM Standards

Think About Power Models

Think about tables and graphs of relations that have equations of the form $y = ax^2$ and $y = ax^3$. What patterns occur in the tables and graphs when a is a positive number? What patterns are there when a is a negative number? Also think about tables and graphs that might have equations in the form $y = x^n$. What patterns do you expect to see if n is an even positive integer? What if n is odd? How are the patterns represented in tables and graphs different for power models, linear models, and exponential models? Write a few sentences to explain your thinking.

Think About Inverse Power Models

Think about equations of the form $y = \dfrac{a}{x}$ and $y = \dfrac{a}{x^2}$. What effect would changes in the value of a have on the graphs of these functions? What patterns would appear in the tables and graphs that could be represented by these equations? Write a few sentences to explain your thinking.

Daily Warm-Ups: Algebra, NCTM Standards

Tyrone's Toothpick Triangles

Tyrone is creating patterns with toothpicks.

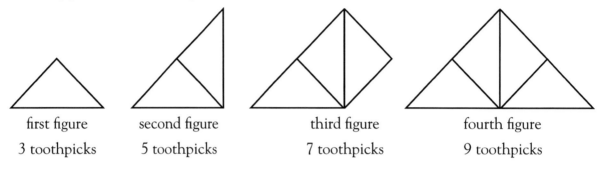

first figure	second figure	third figure	fourth figure
3 toothpicks	5 toothpicks	7 toothpicks	9 toothpicks

1. How many toothpicks are in the eleventh figure in his pattern?

2. Write both an explicit equation and a recursive equation for the number of toothpicks, t_n, for the nth figure in Tyrone's pattern of toothpicks.

3. What would the figure look like in Tyrone's pattern that has exactly 71 toothpicks?

Daily Warm-Ups: Algebra, NCTM Standards

Switching Chips

Rocco is exploring a children's puzzle that involves a strip of 7 squares, 3 white chips, and 3 black chips.

The object of the puzzle is to switch positions of the chips so that all the black chips are on the left and all the white chips are on the right. A chip can move by sliding to an adjacent empty square or by jumping over one chip to land on an empty square. White chips only move right, and black chips only move left.

1. How many moves will it take to switch the chips?

2. Complete the table to show the number of moves it would take for other numbers of chips.

Number of each color	1	2	3	4	5	6	7	8	9	10
Number of moves										

3. A puzzle with n white chips and n black chips will require how many moves to complete the switch?

Daily Warm-Ups: Algebra, NCTM Standards

© 2008 Walch Education

Holiday Sales Prices

As part of a preholiday sales event, a clothing store plans to reduce the price of shirts that usually sell for $30 by 10% each week.

1. The store originally paid $16 per shirt. In what week will the store begin to lose money on the shirts at the sale price?

2. Write a recursive formula for this situation.

3. Do the values of the changing sales prices represent a geometric sequence? Explain why or why not.

Daily Warm-Ups: Algebra, NCTM Standards

Exploring Constant Areas

Juan is exploring the relationship between the widths and lengths of rectangles with a constant area. He states that this relationship should be represented as an inverse function. He has been considering rectangles with an area of 24 in^2 and rectangles with an area of 36 in^2. Do you agree or disagree with Juan's statement? What tables of data might he have? How might he be representing the situation symbolically and graphically?

Granola Inflation

A box of Crunchy Granola is selling for $2.98 at the Healthy Grocery Mart. The inflation rate for granola has been noted to be 4% annually.

1. What do you expect the price to be a year from now?

2. What would you expect the price to have been a year ago?

3. Write an equation that models this information.

4. In what year would you expect the price of a box of Crunchy Granola to be more than $3.50?

Daily Warm-Ups: Algebra, NCTM Standards

Maximizing Potato Profit

Maximilian and Minerva have harvested 10,000 pounds of potatoes this fall. They know that they can sell them now for $0.20 per pound. They also know that the price is likely to go up approximately $0.02 per week if they wait to sell. If they wait to sell, however, it is quite possible that they'll lose approximately 200 pounds of potatoes per week due to spoilage and animals munching. When should they sell to maximize their income for their potatoes?

Depreciating Car Values

Lupita wants to buy a car that will have the best resale value after 3 years. At one dealership, she has found a car she likes that sells for $15,000 and depreciates at a rate of 30% per year. At another dealership, she has found a car that costs $12,000 and depreciates at a rate of 20% per year. Which car will have the greater value in 3 years?

Swimming Pool Bacteria

1. At the local swimming pool, the bacterial count is measured on Monday morning at 8 A.M. and found to be 1,200 bacteria per cubic centimeter. If the count is thought to double every 24 hours, what will the bacteria count be on Thursday afternoon at 2 P.M.?

2. If the safe range of bacteria is less than 200,000 per cubic centimeter, when will this pool need to be treated? Give an equation and a graph with your solution.

Daily Warm-Ups: Algebra, NCTM Standards

Bouncing Ball

A hard rubber ball will rebound to 75% of its height each time it bounces. If the ball is dropped from a height of 200 centimeters, what will the height of each bounce be until the ball rests? Create a table, an equation, and a graph of the ball's bounce over several bounces. On which bounce will the rebound be less than 50 centimeters?

Daily Warm-Ups: Algebra, NCTM Standards

Olympic Pool Border

An Olympic-size swimming pool measures 25 meters by 50 meters. The Harrison City Recreation Department is planning to construct a new pool that will be bordered with a walkway of a uniform width w. Find the possible widths of the walkway if the total area of the walkway is to be greater than 76 square meters, but no more than 400 square meters.

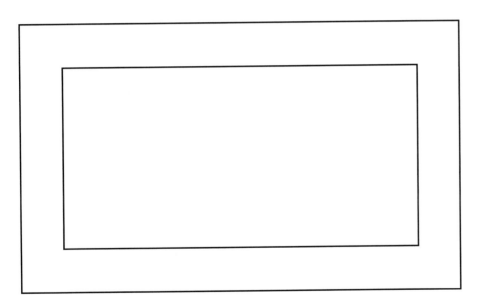

Copy Machine Choices

The Pendleton County School District is trying to decide on a new copier. The purchasing committee has been given quotes on two new machines. One sells for $20,000 and costs $0.02 per copy to operate. The other sells for $17,500, but its operating costs are $0.025 per copy. Which machine would you recommend? Justify your choice with clear mathematics information. How many copies must the school make before the higher price is a reasonable choice?

Sidney's Salary and Rent

Sidney was hired as a first-year teacher to teach high-school mathematics at a starting salary of $35,000. He received an annual salary increase of 5% for each of the next 4 years. Sidney rented an apartment for $950 per month. Each time he renewed his lease over 4 years, the landlord raised the rent by $50.

Using your understanding of arithmetic and geometric sequences, represent Sidney's two situations giving the common difference or common ratio and the recursive formulas. Then explain why you made those choices.

Daily Warm-Ups: Algebra, NCTM Standards

Maximum Area

Sam and Penny have 1,200 feet of fence. They would like to enclose as large a rectangular plot of land as possible. Their farm lies along a river, and they will not need to put up a fence along that side of their land. What is the maximum area that they can get from the length of fence that they have?

Daily Warm-Ups: Algebra, NCTM Standards

Life Expectancy

The table below shows the average number of years a person born in the United States can expect to live based on their year of birth. What do you predict is the life expectancy of a man, a woman, or any person born in 2020? Find a line of best fit for this data. Write an equation for the line you choose. Make your prediction using your graph and your equation.

Year	Female	Male	Both sexes
1930	61.6	58.1	59.7
1940	65.2	60.8	62.9
1950	71.1	65.6	68.2
1960	73.1	66.6	69.7
1970	74.7	67.1	70.8
1975	76.6	68.8	72.6
1980	77.4	70.0	73.7
1985	78.2	71.1	74.7
1990	78.8	71.8	75.4
1995	78.9	72.5	75.8
2000	79.7	74.3	77.0
2005	80.4	75.2	77.9

Daily Warm-Ups: Algebra, NCTM Standards

© 2008 Walch Education

Interpreting Graphs 1

Which graph best represents the given situation? Be prepared to justify your answers.

1. A commuter train pulls into a station and drops off all its passengers.

   a.   b.   c.   d.

2. A child walks to a slide, climbs up to the top, and then slides down.

   a.   b.   c.   d.

Daily Warm-Ups: Algebra, NCTM Standards

Interpreting Graphs II

Which graph best represents the given situation? Be prepared to justify your answers.

1. A little boy swings on a swing at a playground.

   a.   b.   c.   d.

2. A woman walks up a hill at a constant rate and then runs down the other side.

   a.   b.   c.   d.

Daily Warm-Ups: Algebra, NCTM Standards

Ferris Wheel Graphs

Enrique took a ride on a Ferris wheel at the amusement park last weekend. He has sketched these graphs to represent that ride. Which one best represents the idea that he's trying to show? Explain why you chose that graph.

a.

b.

c.

d.

Daily Warm-Ups: Algebra, NCTM Standards

Roller Coaster Ride

Marjorie and Jack love to ride roller coasters and then portray their ride in a graph. Last Saturday, they rode the Mighty Twister and sketched this graph of that experience.

1. What does the pattern of their graph tell you about when they were going down the steepest hill?

2. What was their approximate velocity 1 second into the ride? What was their velocity 8 seconds into the ride?

3. After getting started, when were they going the slowest? The fastest?

4. Describe what was happening to them between 4 and 8 seconds.

Daily Warm-Ups: Algebra, NCTM Standards

U.S. Population

The graph below shows the United States population from 1900 to 2000, as recorded by the U.S. Census Bureau. Use the graph to answer the following questions.

1. What was the rate of change in population from 1900 to 2000? Is this greater or less than the rate of change in the population from 1990 to 2000?

2. Which 10-year time periods have the highest and the lowest rates of change? How did you find these? What factors might have contributed to these rates of change?

3. What do you predict the U.S. population will be in 2010? Explain your reasoning.

Daily Warm-Ups: Algebra, NCTM Standards

Cool It!

The temperature in Celsius of a block of hot metal as it cools is given by the function $T(x) = 95(1 - 0.45)^x + 20$, where x is expressed in hours.

1. What is the starting temperature for this hot metal block?

2. What will the temperature be after 3 hours?

3. What will the temperature be when the block has finished cooling?

Daily Warm-Ups: Algebra, NCTM Standards

Phone Plan Decisions

Aisha's parents are planning to give her a new cell phone for her upcoming birthday and pay for her service for one year. They have asked her to determine which plan is best for her. Best Talk offers service at a monthly basic fee of $20.00 per month and $0.10 per minute for each minute used. Horizon One-Rate has no monthly fee, but charges $0.40 per minute. Create a table, a graph, and equations that model this situation. Then give an argument for why Aisha might choose either plan.

Daily Warm-Ups: Algebra, NCTM Standards

Fair Trade

Land developers sometimes make deals with landowners and trade pieces of land that they need for other plots of land in other places. Suppose you own a square piece of property that the ABC Land Development Corporation would like to have to build a new mall. The company is willing to trade your property for a rectangular piece nearby that is 3 meters longer on one side and 3 meters shorter on the other side. How does this rectangular piece compare to your original square plot? Would this be a fair trade for any side length of your original square? Complete the table and compare the areas. Write expressions for the situation.

Original square		Rectangular plot			Difference in area
Side length	Area	Length	Width	Area	
4	16	7	1	7	9
5					
6					
7					
8					

Daily Warm-Ups: Algebra, NCTM Standards

Used Car Purchase

Mary Ellen has just started a new job and will need a car to drive to work. She stopped by Jolly Joe's dealership and found a car that she really likes. Joe has told her that she can buy the car for a down payment of $2,000 and a monthly payment of $399 for 24 months.

Happy Hank runs a competing dealership nearby. He has a car almost exactly like the one Mary Ellen wants. Hank will sell her the car for a down payment of $1,600 and a monthly payment of $420 for 24 months.

1. Write an equation for each situation that represents the amount that Mary Ellen will have paid up to that point for any month in the 24-month period.

2. What amount will she have paid at the end of 1 year if she buys Jolly Joe's car? What if she buys Happy Hank's car? What will she have paid for each car at the end of 2 years?

3. Which is the better deal?

Daily Warm-Ups: Algebra, NCTM Standards

The Shape-Shifting Square

A square has a side of length s. A new rectangle is created by increasing one dimension by 5 centimeters and by decreasing the other dimension by 4 centimeters.

1. Draw a sketch that represents this situation.

2. Write symbolic expressions that represent the areas of the two figures.

3. For what values of s will the area of the new rectangle be greater than the original square?

4. For what values of s will the area of the new rectangle be less than the original square?

5. For what values of s will the areas of the two figures be equal?

6. Explain how you found your answers.

Daily Warm-Ups: Algebra, NCTM Standards

A Ball Is Thrown

Luis threw a baseball straight up at an initial velocity of 120 feet per second. Luis is 6 feet tall. The height of a ball is a function of the time, t, that it was thrown and can be found using the following equation:

$H(t) = -16t^2 + v_1 t + h_1$

How high did Luis's throw go, and how long did it stay in the air? Create a graph for this situation.

Daily Warm-Ups: Algebra, NCTM Standards

Taking an Interest in Interest

The Wild Cat Valley Savings Bank pays 2.25% annual interest on an individual savings account and compounds monthly. The amount of money after n months is given by $A = P(1 + \frac{0.0225}{12})^n$, where P is the initial amount deposited. Francis deposited $1,000 to start her savings.

1. What will Francis's savings be after 5 years?

2. How long will it take for the amount in her account to double if she does not withdraw any funds?

3. If the annual interest rate was doubled, how long would it take her savings to double?

Daily Warm-Ups: Algebra, NCTM Standards

Oops, Decayed Equation!

During their investigation of radioactive decay, a research group discovered that the equation $y = 200(1 - 0.18)^x$ fits their data. Unfortunately, one member of the group spilled coffee on the table and made the data unreadable. Reconstruct what their table might have looked like using the chart below. Explain what is being represented in the equation by 200 and 0.18.

x	y
0	
1	
2	
3	
4	
5	
6	

Daily Warm-Ups: Algebra, NCTM Standards

Water Balloon Experiments

Mr. Zeno's algebra class is conducting experiments with water balloons from the roof of the Pendleton County High School. Clarissa is using the formula $h = -4.9t^2 + h_1$, where h is the height of a falling object. She knows that t represents time in seconds and h_1 is the starting height of the object. Her group has found that the height of the high school is 30 meters. Graph Clarissa's information. Using your graph, find out when the balloon will be 10 meters from the ground.

Daily Warm-Ups: Algebra, NCTM Standards

Bacteria Population Decay

Marcus and his research group are studying a bacteria population. They believe that the equation $p(t) = 800(1-0.093)^t$ models the change in the bacteria population over time in hours.

1. How many bacteria were there when Marcus began his study?

2. Sketch the graph that models this bacteria population. What is a good window for this situation? Is the graph a function? Why or why not?

3. Use your model to determine the half-life of this bacteria population. How long will it take to decrease the population to half its original size?

Daily Warm-Ups: Algebra, NCTM Standards

Tennis Ball Trauma

A tennis ball has been dropped from the top of a tall building. The ball's height in meters t seconds after it is released can be represented by $h(t) = -4.9t^2 + 150$.

1. Find $h(3)$ and explain what this represents in the situation described.

2. How much time will elapse (to the nearest .01 second) until the ball is at or less than 25 meters above the ground?

3. When will the tennis ball hit the ground?

Daily Warm-Ups: Algebra, NCTM Standards

Modeling Tidal Cycles

Melinda has created a trigonometric equation. She believes the equation models the height of water on the pylons of her father's dock during the tide cycle. Her equation is $h = 7.5\cos\left(\dfrac{2\pi(t-3)}{12}\right) + 16$, where t represents the time in hours since midnight and h represents the height in feet. Melinda's father wants to launch his boat when the water is approximately 12 feet high on the pylons. He thinks that this will happen around 8:00 A.M. What do you think about this prediction? Using Melinda's equation model, predict the first time when the height of the water will be approximately 12 feet on the dock pylons. When will the water be that height again?

Daily Warm-Ups: Algebra, NCTM Standards

Part 2: Analyzing Change in Various Contexts

National Council of Teachers of Mathematics: "Instructional programs from pre-kindergarten through grade 12 should enable all students to analyze change in various contexts."

Expectations

Grades 6–8:
- Use graphs to analyze the nature of changes in quantities in linear relationships.

Grades 9–12:
- Approximate and interpret rates of change from graphical and numerical data.

Parachuting Down

Cecelia took her first parachute jump lesson last weekend. Her instructor gave her the graph below that shows her change in altitude in meters during a 2-second interval. Use the graph to answer the questions that follow.

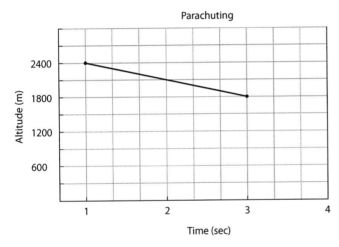

1. What is the slope of the line segment?
2. Estimate Cecelia's average rate of change in altitude in meters per second.
3. Give the domain and range for this graph.

Daily Warm-Ups: Algebra, NCTM Standards

Ping-Pong Prices

Lily found the price of ping-pong balls listed on the Internet at $4.75 for a package of 6 balls. Shipping and handling was listed at $1.00 per package.

1. Write an equation that represents the total cost for different numbers of packages of ping-pong balls.

2. Sketch a graph of this relationship.

3. If you shift your graph up a value of $.50, does this mean the price per package increased, or the shipping price increased?

4. Write a new equation for the situation in question 3.

Daily Warm-Ups: Algebra, NCTM Standards

Graphing People Over Time

Think about each situation below. Then sketch a graph to represent the situation over a 24-hour period of time. Label each graph carefully using the horizontal axis to represent the time of day. Be prepared to explain and justify your choices.

1. the number of people at a very popular pizza restaurant on a Saturday

2. the number of people in a school building on a weekday in September

3. the number of people at a sports stadium on the day of a big game

4. the number of people at a movie theater on a weekend day

Daily Warm-Ups: Algebra, NCTM Standards

DVD Rentals

Rosa's parents just bought a new DVD player for the family. Rosa's mom asked her to research rental prices for DVDs online. More Movies has a yearly membership package. Rosa found the following table of prices on the company's web site.

More Movies DVD Rental-Membership Packages

Number of videos rented	0	5	10	15	20	25	30
Total cost	$30	$35	$40	$45	$50	$55	$60

On the Deluxe DVD Rentals web site, Rosa found that there was no membership package offered. She made the graph on the right to show how the cost at Deluxe DVD Rentals is related to the number of videos rented.

If both rental services have a comparable selection of DVDs and Rosa's family will watch on average about two movies a month, which service should they choose? Explain how Rosa's family might decide which service to use. Using the information about each service, describe the pattern of change relating the number of DVDs rented to the total cost.

Camping Costs

Dan and Julia are planning a camping trip for their school's Outdoor Adventure Club. They went online and found the following information for Hamden Hills State Park. Use the information to answer the questions below.

Number of campsites needed	1	2	3	4	5	6	7
Total campground fee	$17.50	$35.00	$52.50	$70.00	$87.50	$105.00	$122.50

1. Make a coordinate graph of these data. Would it make sense to connect the points on the graph? Why or why not?

2. Using the table, describe the pattern of change in the total fee as the number of campsites needed increases. How is this pattern shown in your graph?

Daily Warm-Ups: Algebra, NCTM Standards

© 2008 Walch Education

Renting Canoes

Dan and Julia have reservations for 10 students for the Outdoor Adventure Club's annual trip. This year they are planning a canoe camping trip. The Otter Mountain River Livery rents canoe and camping gear for $19 per person. Dan and Julia expect no more than 50 students in all to go on the trip. Using increments of 10 campers, make a table showing the total rental charge for 10 to 50 campers. Then make a coordinate graph of these data.

Counting Step-Ups

Alejandro has collected data on the number of step-ups his friend Jalesia can do in 2 minutes. The graph shows their data.

1. What are the two variables represented on the graph?

2. Make a table of Alejandro's data. Describe how the number of step-ups changes for each 10-second interval as time increases.

3. From Alejandro's data, estimate the number of step-ups Jalesia did in 25 seconds and in 65 seconds. Explain how you made those estimates.

Daily Warm-Ups: Algebra, NCTM Standards

Thinking About Variables, Graphs, and Tables

1. Think about a situation in which variable y depends on variable x. (For example, y might be profit and x the number of items sold.) If y increases as x increases, how would this appear in a table? How would it appear in a graph?

2. Now think about a situation in which variable y decreases as variable x increases. (For example, y might be the amount of gasoline in your car on a trip and x the time you have been traveling.) How would this be indicated in a table? In a graph?

3. In a coordinate graph of two related variables, when do the points lie in a straight line?

4. In a coordinate graph of two related variables, when is it appropriate to connect the points?

Daily Warm-Ups: Algebra, NCTM Standards

Grow Baby!

The average growth weight in pounds of a baby born in the United States is a function of the baby's age in months. Look at the sample data provided and graph the information in an appropriate window. Then write two sentences about the graph and the relationship between the two variables.

Age in months	Weight in pounds
0	7
3	13
6	17
9	20
12	22
15	24
18	25
21	26
24	27

Daily Warm-Ups: Algebra, NCTM Standards

© 2008 Walch Education

Thinking About Analyzing Change

Think about the graph below. It represents the speed over time for two different automobiles.

Speed of Two Cars

$T = 0$ hour $\qquad\qquad\qquad T = 1$ hour

If the cars started from the same place at the same time, what is the relationship between the two cars after 1 hour? Discuss the velocity and acceleration of each car at $T = 1$ hour. Write a few sentences that explain your thinking.

Motion Mystery

Three students are learning to use a motion detector system such as the CBL or CBR units available for graphing calculators. They all walked a short distance and created graphs on their calculators, which recorded the data. Now the calculators are mixed up. The students are no longer certain which calculator recorded which student's data. The graphs are shown below. To help the students identify the calculators, describe in detail what you think each of the graphs reveals about the walk that was recorded.

a.

b.

c.
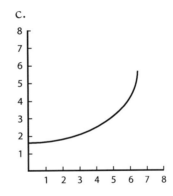

Graphing Inequalities 1

Graph the region indicated by each pair of inequalities below.

1. $y \leq -2x + 7$ and $y \geq 4x + 3$
2. $-3x + 2y > 6$ and $y \leq -2x + 5$

Daily Warm-Ups: Algebra, NCTM Standards

Graphing Inequalities II

Describe the region represented on each graph below using appropriate inequalities.

1.

2.

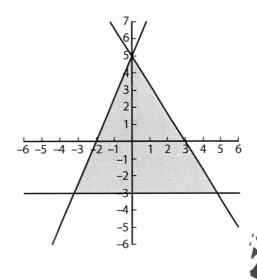

Daily Warm-Ups: Algebra, NCTM Standards

Imagining the Graphs of Sequences

Look at the recursive formulas below. Without actually graphing them, imagine the graphs of the sequences generated by these formulas. Then describe each graph using one or more of the following terms: arithmetic, geometric, decreasing, increasing, linear, and nonlinear. Be prepared to justify your thinking.

1. $t_0 = 60$, $t_n = t_{n-1} - 10$, for $n \geq 1$

2. $t_0 = 1{,}000$, $t_n = 0.7t_{n-1} + 100$, for $n \geq 1$

3. $t_0 = 7.5$, $t_n = 0.2 + t_{n-1}$, for $n \geq 1$

4. $t_0 = 98$, $t_n = (1 + 0.25)t_{n-1} - 10$, for $n \geq 1$

5. $t_0 = 75$, $t_n = t_{n-1} \times 1.75$, for $n \geq 1$

Braking/Stopping Distances

Use a graphing calculator or graph paper to graph the relationship between miles per hour and stopping distance using the data from the table below. What questions come to mind as you look at your graph? Make a prediction for the stopping distance at 80 miles per hour. Explain and justify your prediction.

Miles per hour	Stopping distance
10	27
15	44
20	63
25	85
30	109
35	136
40	164
45	196
50	229
55	265
60	304
65	345

Daily Warm-Ups: Algebra, NCTM Standards

Part 3: Using Algebraic Symbols I

National Council of Teachers of Mathematics: "Instructional programs from pre-kindergarten through grade 12 should enable all students to represent and analyze mathematical situations and structures using algebraic symbols."

Expectations

Grades 6–8:
- Develop an initial conceptual understanding of different uses of variables.
- Explore relationships between symbolic expressions and graphs of lines, paying particular attention to the meaning of intercept and slope.
- Use symbolic algebra to represent situations and to solve problems, especially those that involve linear relationships.
- Recognize and generate equivalent forms for simple algebraic expressions and solve linear equations.

Hedwig's Hexagons

Hedwig is using toothpicks to build hexagon patterns. The first three are pictured. How do you think her pattern will continue? She has begun to collect information in a table in order to explore the relationships among the number of toothpicks, the number of hexagons, and the length of the outside perimeter of the figure. Complete the table for her.

Number of toothpicks			
Number of hexagons	1	2	
Perimeter of figure		10	

Now write an equation for the relationship between the number of hexagons in each figure and the perimeter of the figure. Write another equation for the relationship between the number of hexagons and the number of toothpicks.

Daily Warm-Ups: Algebra, NCTM Standards

Evaluating Simple Expressions 1

Evaluate each expression below for the given value of *x*.

1. $4.5x - 12$ for $x = 2$

2. $-5 - x$ for $x = \dfrac{1}{2}$

3. $2x^2$ for $x = 8$

4. $x^2 + 2.5$ for $x = 1.5$

5. $(x - 2.5)(x + 35)$ for $x = 2.5$

6. $x(27 - x)$ for $x = 3$

7. $\dfrac{49}{x^2}$ for $x = -7$

8. $5x^2 + 3x - 10$ for $x = 2$

9. $\dfrac{x}{4}$ for $x = 8$

10. $.5x^2 + 2x - 20$ for $x = 10$

Daily Warm-Ups: Algebra, NCTM Standards

Evaluating Simple Expressions II

Evaluate each expression below for the given value of x.

1. $42 - 3x$ for $x = 6$

2. $13 - 3x^2$ for $x = 1$

3. $4x^2 + 13$ for $x = -10$

4. $6x^2 + x - 2$ for $x = 2$

5. $(x - 2.5)(x + 5)$ for $x = 0$

6. $x(27 - x)$ for $x = -3$

7. $\dfrac{3(15 - x)}{2x}$ for $x = 10$

8. $8x - 3x(6 - x)$ for $x = 0$

9. $\dfrac{x^2}{4}(x + 8)$ for $x = -8$

10. $-2x^2 + 5x + 12$ for $x = -4$

Daily Warm-Ups: Algebra, NCTM Standards

Thinking Around the Box

The rectangular box shown below has a front and back length of L meters. The side lengths are W meters. The height is H meters.

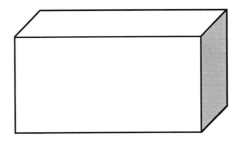

1. Write two equations that represent the sum, S, of all the edges of the rectangular box.

2. Write two equations that represent the surface area, A, of the rectangular box.

Daily Warm-Ups: Algebra, NCTM Standards

Perimeter Problem

Use the figure below to answer the questions that follow.

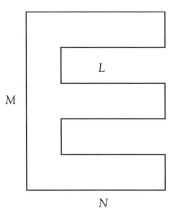

1. If $M = 5$ meters, $N = 4$ meters, and $L = 2.5$ meters, what is the perimeter of the figure?

2. Using the variables only, write two or more expressions that could represent the perimeter of the figure. Show that your expressions are equivalent.

3. Is there another way to represent the perimeter of the figure?

Daily Warm-Ups: Algebra, NCTM Standards

Solving Equations 1

Tucker is tutoring his cousin James in pre-algebra. He is trying to explain different strategies for solving an equation such as $53 = 17 - 4x$. What steps or approaches should he recommend to James? Explain your ideas carefully.

Daily Warm-Ups: Algebra, NCTM Standards

Solving Equations II

Solve each equation below symbolically. Be prepared to justify your answers.

1. $3x + 7 = 19$

2. $7 + 4x = 31$

3. $26 - 3x = 4x + 19$

4. $x^2 - 3.5x = 0$

5. $13x - x^2 = 0$

6. $x^2 - 10x = 5x^2 - 2x$

7. $4x(x - 5) = 0$

8. $3.5(x + 2) + 2(x + 2) = 0$

9. $5.5x = x - 9$

10. $49x - 7x^2 = 0$

Daily Warm-Ups: Algebra, NCTM Standards

Simplifying Expressions

Write at least two more expressions that are equivalent to each given expression below.

1. $8(x - 5)$

2. $x(5x - 6) + 13x - 10$

3. $4(x + 5) - 3(2 - 4x)$

4. $2.5(12 - 2x) + 5(x + 1)$

5. $(3x^2 + 5x + 8) - (8x^2 + 2x - 5)$

Daily Warm-Ups: Algebra, NCTM Standards

Slope-Intercept Equations I

Simplify the following equations so that you could enter them more easily into a graphing calculator. Give the slope and y-intercept for each without graphing it.

1. $y - 3(x - 7) = 5x$

2. $y = -2 + (x + 1)$

3. $y + 4 = \dfrac{5}{3}(x + 6)$

4. $y = 28 + 2.5(x - 6)$

5. $y = 13.2(x - 20) + 125.6$

Daily Warm-Ups: Algebra, NCTM Standards

Slope-Intercept Equations II

Find an equation written in slope-intercept form that satisfies each condition below.

1. a line whose slope is -3 and passes through the point $(2, 5)$

2. a line whose slope is $\frac{-2}{3}$ and passes through the point $(4, -1)$

3. a line that passes through the points $(2, 6)$ and $(6, 1)$

4. a line that passes through the points $(-2, 3)$ and $(4, -3)$

5. a line that passes through the points $(-1, -1)$ and $(-5, -5)$

Daily Warm-Ups: Algebra, NCTM Standards

© 2008 Walch Education

Up and Down the Line I

Write an equation for the line pictured in each graph below.

1.

2.

3.

4.

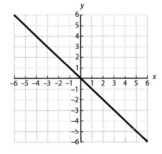

Daily Warm-Ups: Algebra, NCTM Standards

Up and Down the Line II

Write a linear equation for each condition below.

1.

x	−1	0	1	2	3
y	1	3	5	7	9

2. a line whose slope is −3 and y-intercept is 5

3. a line that passes through the points (2, 5) and (5, 6)

4. a line that passes through the point (3, 7) and has a slope of $\frac{2}{3}$

Daily Warm-Ups: Algebra, NCTM Standards

Butler Bake Sale

The Butler High School sophomore class is planning a bake sale as a fall fund-raiser. Luis is chairing the planning committee. He gave a brief survey to determine what price should be charged for each brownie. He predicts from his results that a price of $0.50 per brownie will result in 200 brownies sold. He also predicts that a price of $1.00 per brownie will result in about 50 brownies sold. He is assuming that the relationship between the brownie price and the number sold is a linear relationship.

1. Write an equation for the relationship that Luis has predicted between the cost and the number of brownies sold.

2. Find the slope and y-intercept. Then explain what these mean in the context of Luis's information.

3. What if the committee decides to charge $0.70 per brownie? How many can they expect to sell?

4. Luis has taken an inventory and found that the class has 300 brownies to sell. Using your equation, what would be an appropriate price to charge for each brownie?

Looking for Lines I

Explain how you can find the equation of a line if you know the information below. Use examples to explain your thinking.

1. the slope and y-intercept

2. two points on the line

3. the slope of the line and a point that is on the line, but is not the y-intercept

Daily Warm-Ups: Algebra, NCTM Standards

Looking for Lines II

Linear patterns can be recognized in several different representations of a relationship between two sets of quantities. Sometimes we look for the relationship in graphs, tables, or symbolic rules. Describe how you can tell whether a situation is or can be a linear model by looking at the information below.

1. a scatter plot of the data

2. a table of the values

3. the equation of the relationship

4. a description of the problem

Daily Warm-Ups: Algebra, NCTM Standards

Thinking About Equations of Linear Models

Think about ways to find an equation of the form $y = mx + b$ or $y = a + bx$ from a table of data or a graph of the points. How can you find the equation if you know the slope and y-intercept? How can you find the graph by looking for the rate of change and other values from the table? How can you find the equation of the line if the slope and y-intercept are not given? Write a few sentences to explain your thinking.

Thinking About Change and Intercepts in Linear Models

Think about linear models in various forms. How can you see or find the rate of change in a linear model from a table of values? How would you find it from the graph? How does it appear in the equation? How would you determine the y-intercept in those three situations? Write a few sentences to explain your thinking.

Simplifying Algebraic Fractions

Sometimes it is helpful to express an algebraic fraction in a form that is simpler than the way that it appears. If possible, express each fraction below in a simpler form. If the fraction cannot be expressed in a simpler form, write *not possible*.

1. $\dfrac{x^2 - 4}{x - 2}$

2. $\dfrac{n^2 + 1}{n + 1}$

3. $\dfrac{b^2 + b}{b + 1}$

4. $\dfrac{f + 15}{f^2 + 15f}$

5. $\dfrac{x^2 + 7x + 10}{x + 2}$

6. $\dfrac{n^2 - 8n + 16}{n - 4}$

Daily Warm-Ups: Algebra, NCTM Standards

Thinking About Linear Models

Imagine that two of your classmates have proposed two different linear models for a table of data that you are investigating. This means that they have given you two different lines and two different equations for the situation. How can you compare the two models and the data to find out which is the best model for the situation? Write a few sentences to explain your thinking.

Chirping Crickets

Did you know that you can estimate the temperature outside by counting the number of cricket chirps you hear? One method to estimate the temperature in degrees Fahrenheit is to count the number of chirps in 15 seconds and then add 37. This will give you the approximate temperature outside.

1. Write a mathematical sentence that describes the number of cricket chirps per 15 seconds (C) as a function of the temperature (T) in Fahrenheit.

2. Graph this function.

3. What do you think are the maximum and minimum values of this function?

Daily Warm-Ups: Algebra, NCTM Standards

Freezing or Boiling?

Lucas knows that the relationship between Celsius and Fahrenheit temperatures is a linear one. However, he often forgets the equation. He remembers that at freezing, the temperatures are (0°C, 32°F). At boiling, the respective temperatures are (100°C, 212°F). Explain to Lucas how he can find this equation knowing that the relationship is a linear one.

The Seesaw Problem

1. Eric and his little sister Amber enjoy playing on the seesaw at the playground. Amber weighs 65 pounds. Eric and Amber balance perfectly when Amber sits about 4 feet from the center and Eric sits about $2\frac{1}{2}$ feet from the center. About how much does Eric weigh?

2. Their little cousin Aleah joins them and sits with Amber. Can Eric balance the seesaw with both Amber and Aleah on one side, if Aleah weighs about the same as Amber? If so, where should he sit? If not, why not?

Daily Warm-Ups: Algebra, NCTM Standards

Weekly Pay Representations

Amanda has a part-time job working for a local company selling wireless phone subscriptions. She is paid $120 per week plus $25 for each subscription that she sells.

1. Complete the table below, showing Amanda's weekly pay as a function of the number of subscriptions that she sells.

Number of subscriptions sold	Weekly pay (dollars)
1	
2	
3	
6	
10	
12	
15	

2. Using variables, write two rules (equations), one for the weekly pay in terms of her sales, and one for her sales in terms of her pay.

3. Amanda wants to earn at least $600 next week. How many subscriptions must she sell?

Daily Warm-Ups: Algebra, NCTM Standards

Symbol Sense: Original Costs

Think of a real-world situation that might result in each symbolic sentence below. Solve the sentences for C.

1. C − .25C = $37.49

2. $28.00 + .20C = C

Babysitting Bonus

This year, Zachary has been babysitting his young cousins after school for $70 a month. His uncle also gave him an extra bonus of $100 for his excellent work. Since school started, Zachary has earned more than $500. How many months ago did school start? Write an inequality that represents this situation. Solve it showing all your work.

Part 4: Using Algebraic Symbols II

National Council of Teachers of Mathematics: "Instructional programs from pre-kindergarten through grade 12 should enable all students to represent and analyze mathematical situations and structures using algebraic symbols."

Expectations

Grades 9–12:
- Understand the meaning of equivalent forms of expressions, equations, inequalities, and relations.
- Write equivalent forms of equations, inequalities, and systems of equations and solve them with fluency—mentally or with paper and pencil in simple cases and using technology in all cases.
- Use symbolic algebra to represent and explain mathematical relationships.
- Use a variety of symbolic representations, including recursive and parametric equations, for functions and relations.
- Judge the meaning, utility, and reasonableness of the results of symbol manipulations, including those carried out by technology.

Border Tiles

Landscapers often use square tiles as borders for garden plots and pools. The drawing below represents a square pool for goldfish surrounded by 1-foot square tiles.

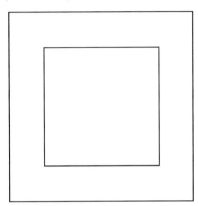

Maria thinks that the total number of tiles around the outside of the pool could be represented by $(n + 2) + (n + 2) + n + n$. Carlos thinks that he can represent the total number of tiles by $4(n + 1)$. What do you think about Carlos and Maria's representations? How do you think they would explain their thinking? Can you show if their expressions are equivalent or not?

Daily Warm-Ups: Algebra, NCTM Standards

Perimeter Expressions

Rajah is working on her math homework. She has been given the sketch below, where r represents both the radius of the semicircle and the height of the rectangle that the semicircle is attached to. She is trying to decide on an expression that will represent the perimeter of the figure. She has written out the expressions below. Do they correctly represent the perimeter of the shape? Explain why or why not for each.

1. $r(2 + \pi)$
2. $4r + \pi r$
3. $r + 2r + r + 2\pi r$
4. $r(4 + \pi)$
5. $2r + \pi r$

Daily Warm-Ups: Algebra, NCTM Standards

Cylinder Expressions

The cylinder below has a base area of 45 cm². It is partially filled with liquid up to a height of x cm. The height from the top of the liquid to the top of the cylinder is represented by y cm. The total height is represented by h cm. Which expressions below properly represent the total volume of the cylinder?

a. $45(h - x)$

b. $45(x + y)$

c. $45x + 45y$

d. $45xy$

Daily Warm-Ups: Algebra, NCTM Standards

© 2008 Walch Education

A Hundred Puzzling Fractions

Making number patterns and puzzles is one of Keiko's favorite activities. Here are two of her most recent creations. Solve each puzzle, and then explain how you solved it.

1. $(1-\frac{1}{2})(1-\frac{1}{3})(1-\frac{1}{4})\ldots(1-\frac{1}{98})(1-\frac{1}{99})(1-\frac{1}{100}) = ?$

2. $\frac{1}{1\cdot 2}+\frac{1}{2\cdot 3}+\frac{1}{3\cdot 4}+\ldots\frac{1}{98\cdot 99}+\frac{1}{99\cdot 100} = ?$

The Profit Is in the Symbols

Molly and Joe are co-chairs of their school's concert committee. They are planning to have a rock band perform at the school. They have been told that the equation $P(t) = -75t^2 + 1500t - 4800$ reflects their potential profit, P, where t represents the price of a ticket. Molly rewrites the equation as $P(t) = (300 - 75t)(t - 16)$. Joe rewrites it as $P(t) = t(-75t + 1500) - 4800$.

1. Are these equations equivalent? What different ideas were Molly and Joe thinking about when they wrote their equations? What might they learn from the equations?

2. Graph the original equation. What are the break-even values for ticket prices for this concert? What is the fixed cost of the concert? According to this equation model, how would the number of tickets sold change if the price is increased? What might be the best price to charge for tickets? Explain your thinking.

Daily Warm-Ups: Algebra, NCTM Standards

Manipulating Matrices

Perform the indicated operations on the given matrices and scalars without using a calculator.

$s_1 = 4$, $s_2 = -3$ $[A] = \begin{bmatrix} 3 & 6 \\ 4 & 2 \end{bmatrix}$ $[B] = \begin{bmatrix} -2 & 4 \\ 5 & -1 \end{bmatrix}$ $[C] = \begin{bmatrix} -2 & 4 & 3 \end{bmatrix}$

$[D] = \begin{bmatrix} 3 & -2 \end{bmatrix}$ $[E] = \begin{bmatrix} -1 & -5 \\ 3 & -4 \\ 2 & 3 \end{bmatrix}$ $[F] = \begin{bmatrix} 3 & -2 & 4 \\ -1 & 5 & -3 \end{bmatrix}$

1. $[A] + [B]$
2. $s_1 \bullet [B]$
3. $[F] \bullet [B]$
4. $[B] \bullet [B]$
5. $[F] \bullet [E]$

Daily Warm-Ups: Algebra, NCTM Standards

More Manipulating Matrices

Perform the indicated operations on the given matrices and scalars without using a calculator.

$s_1 = 4$, $s_2 = -3$ $\quad\quad [A] = \begin{bmatrix} 3 & 6 \\ 4 & 2 \end{bmatrix} \quad\quad [B] = \begin{bmatrix} -2 & 4 \\ 5 & -1 \end{bmatrix} \quad\quad [C] = [-2 \ \ 4 \ \ 3]$

$[D] = [3 \ \ -2] \quad\quad [E] = \begin{bmatrix} -1 & -5 \\ 3 & -4 \\ 2 & 3 \end{bmatrix} \quad\quad [F] = \begin{bmatrix} 3 & -2 & 4 \\ -1 & 5 & -3 \end{bmatrix}$

1. $s_1 \cdot [A] - s_2 [B]$
2. $[D] \cdot [B]$
3. $[B] \cdot [C]$
4. $[E] \cdot [F]$

Daily Warm-Ups: Algebra, NCTM Standards

Thinking About Using Matrices

Think about a real-world context for creating matrices. What are some advantages and disadvantages of organizing and representing the information in a matrix format? Explain how data might be represented in matrices in more than one way. Write a few sentences to explain your thinking.

Thinking About Multiplying Matrices

Think about the process of multiplying matrices. Describe in detail how to multiply two matrices and any limitations there might be. Does the order that you multiply matrices matter? Write a few sentences to explain your thinking.

Thinking About Matrix Operations

Think about and create some sample matrices. Describe the conditions when matrices can be added and multiplied. What are the dimensions of the sums and products of matrices? Define and give examples of an inverse of a matrix, an opposite of a matrix, an identity matrix, and a matrix that will not have an inverse.

Refreshing Radical Ideas

In your work with algebraic expressions and equations, you will encounter real and complex numbers of many forms. Sometimes it is helpful to rewrite them in equivalent forms. Convert each expression below into an equivalent form in which the denominator contains a simple rational term, and all possible like terms have been combined.

1. $\dfrac{6}{\sqrt{3}}$

2. $\dfrac{2}{\sqrt{6}}$

3. $\dfrac{\sqrt{15}}{\sqrt{3}}$

4. $3\sqrt{2} + 2\sqrt{32}$

5. $\dfrac{21\sqrt{150}}{7\sqrt{2}}$

6. $\dfrac{2 - 3i}{3 + 2i}$

7. $\dfrac{a + bi}{a - bi}$

Daily Warm-Ups: Algebra, NCTM Standards

Exponential Expressions

The following exponential expressions can be rewritten in other forms. Use your understanding of how exponents behave to change each expression to an equivalent one.

1. x^{-n}

2. $\dfrac{y^r}{y^s}$

3. $(x^a)^b$

4. $(ab)^n$

5. $\left(\dfrac{d}{t}\right)^n$

6. $\dfrac{1}{x^n}$

7. $n^p \bullet n^q$

8. $y^{\frac{1}{n}}$

Daily Warm-Ups: Algebra, NCTM Standards

Nathan's Number Puzzles 1

Nathan likes to make up puzzles about integers. Some of his recent puzzles are below. Write symbolic sentences that represent Nathan's puzzles. Then solve each puzzle.

1. Two numbers have a sum of 10. If you add the first number to twice the second number, the result is 8. What are the numbers?

2. One number is twice as large as a second number. The sum of the two numbers is 15. What are the numbers?

3. The first number minus the second number is 2. Twice the first number minus twice the second number is 4. What are the numbers?

Daily Warm-Ups: Algebra, NCTM Standards

Nathan's Number Puzzles II

Nathan has written pairs of linear sentences in symbols. Now he's trying to create word puzzles to go with them. Help him by creating a word puzzle for each pair of linear sentences below. Then find the numbers that fit his equations.

1. $x + 3y = 5$
   $3x + y = 7$

2. $x + 3y = 10$
   $3x + 14 = 2y$

3. $x + 2y = 7$
   $x + 2y = 17$

Systems of Linear Equations 1

Marika has asked you to help her understand how to solve systems of equations. Solve each system of equations below using a different strategy. Then explain to Marika why you chose that strategy for that system. Which are best solved by substitution? Which might be easily graphed? Which could be solved by elimination?

1. $y = x - 1$
   $3x - 4y = 8$

2. $3x + 2y = -10$
   $2x + 3y = 0$

3. $x + y = -10$
   $.5x + 1.5y = 5$

4. $3x - 2y = 6$
   $-2x + 3y = 0$

Daily Warm-Ups: Algebra, NCTM Standards

Systems of Linear Equations II

Amal is learning about systems of linear equations. He has come up with some questions regarding equations such as $y = 3x + 8$ and $y = -5x + 11$. Help him by answering the questions below.

1. What is the objective for solving a system of equations such as the one given?

2. How could you find the solution by graphing?

3. How could you find the equation by using tables of values?

4. How might you find the solution for the system without a graph or a table?

5. What would you look for in the table, the graph, or in the equations themselves that would indicate that a system of equations has no solution?

Daily Warm-Ups: Algebra, NCTM Standards

Factoring Quadratics

Sometimes quadratic equations contain hidden information that is revealed when the equations are written in factored form. Change each quadratic equation below to an equivalent equation in factored form.

1. $y = -16t^2 - 48t$

2. $y = 5x^2 - 10x$

3. $y = 3(x - 1) + x(x - 1)$

4. $y = 52x^2 - 13$

5. $y = x^2 + 5x + 4$

6. $y = x(2x + 5) + x(x - 10)$

Daily Warm-Ups: Algebra, NCTM Standards

Solving Systems Using Matrices

Theresa has been given the following system of linear equations that models a situation that she is investigating. Create the matrix equation that would also model the situation. Then solve for the two variables.

$$6b + 4s = 1750$$
$$b + s = 400$$

Daily Warm-Ups: Algebra, NCTM Standards

Quincy's Quiz

Quincy has been confused about simplifying algebraic expressions. His brother Eli prepared this quiz for him to practice. Look at Quincy's responses below. If the response is correct, write *correct*. If it is incorrect, write the correct answer.

1. $4y - 5y$ $-y$
2. $3x + 2x$ $5x^2$
3. $2n^2 + 4n^2$ $6n^4$
4. $2x \cdot 3y$ $5xy$
5. $2a - (a - b)$ $a - b$
6. $\dfrac{n+5}{n}$ 5
7. $4 - (n + 7)$ $-3 - n$
8. $\dfrac{2x^6}{x^3}$ $2x^2$
9. $(2xy)^2$ $4x^2y^2$
10. $2a + 3b + 3a + 4b$ $6a^2 + 12b^2$

Daily Warm-Ups: Algebra, NCTM Standards

Bicycle Sales Matrices

The first matrix below represents the combined quarterly sales of three brands of mountain bikes and hybrid bikes at Bikes Unlimited in three large American cities. The second matrix shows the quarterly sales of hybrid bikes.

Combined Sales

$$\begin{array}{c} \\ \text{Los Angeles} \\ \text{Boston} \\ \text{Chicago} \end{array} \begin{array}{ccc} \text{Trek} & \text{K2} & \text{Schwinn} \\ \begin{bmatrix} 150 & 95 & 105 \\ 75 & 75 & 90 \\ 85 & 110 & 175 \end{bmatrix} \end{array}$$

Hybrid Sales

$$\begin{array}{c} \\ \text{Los Angeles} \\ \text{Boston} \\ \text{Chicago} \end{array} \begin{array}{ccc} \text{Trek} & \text{K2} & \text{Schwinn} \\ \begin{bmatrix} 50 & 40 & 70 \\ 30 & 35 & 50 \\ 55 & 50 & 150 \end{bmatrix} \end{array}$$

Find the matrix that represents the quarterly sales of mountain bikes. Find the matrix that represents the quarterly sales of mountain bikes and hybrid bikes in Boston for the three brands. Then construct one matrix that shows the total sales for all three brands in all three cities combined. Label the rows of your matrix with the brand names. Label the columns with the terms *hybrid* or *mountain*. What matrix operations did you use to construct each of these matrices?

Daily Warm-Ups: Algebra, NCTM Standards

Dancing to the Beat of Logarithms

Use symbolic reasoning and the properties of exponents and logarithms to simplify the expression or solve the equation for x.

1. $5 = 2(1.3)^x$

2. $60 = 5(1 + \frac{.025}{12})^x$

3. $4\log 8 + 3\log 5 - 3\log 2 = ?$

4. $\log x^3 + \log x^2 + 3\log x = ?$

5. $2^{2x+2} = 8^{x+2}$

Daily Warm-Ups: Algebra, NCTM Standards

Messy Equations

Elizabeth was going to complete her algebra assignment using her graphing calculator. Her older sister Carolyn decided to change the equations that she had copied into her notebook to tease her. Carolyn tells her that the equations are still equivalent to what her teacher had given her. Help Elizabeth by making each equation below quicker and easier to type into her calculator.

1. $y = 2x^{-1}$

2. $y = (2x)^{-1}$

3. $y = \dfrac{7.2x^7}{3.6x^4}$

4. $y = (-4)^0$

5. $y = x^{-4.2} \, x^{6.2}$

6. $y = 3^3 \cdot 3^2$

7. $y = \dfrac{1}{5^{-4}}$

Sums of Consecutive Integers

Consider the following:

$$7 = 3 + 4$$
$$9 = 2 + 3 + 4$$
$$22 = 4 + 5 + 6 + 7$$

These are examples of numbers that can be represented as the sum of consecutive integers. Investigate the numbers from 1 to 35. Find all the ways you can to write each number as the sum of two or more consecutive integers. Look for patterns in your work. Can you predict what the results will be for the number 45? For 64? For 80?

Daily Warm-Ups: Algebra, NCTM Standards

Percent Increase or Decrease

Look at the sequences below. For each, tell whether there is a growth or decay, identify the common ratio, and give the percent increase or decrease.

1. 43, 129, 387, 1,161, . . .

2. 90, 99, 108.9, 119.79, . . .

3. 1,800, 1,080, 648, 388.8, . . .

4. 17.8, 3.56, 0.712, 0.1424, . . .

5. 375, 142.5, 54.15, 20.577, . . .

Daily Warm-Ups: Algebra, NCTM Standards

Real or Imagined?

Write a description of how you would explain to a friend that $x^2 + 9 = 0$ has no real roots.

Daily Warm-Ups: Algebra, NCTM Standards

Rolling Along

Did you know that you can calculate the speed (S) of a car by knowing the size of its tires and the revolutions per minute (RPM) that the tires make as they rotate? Think about a car that has tires that are 2.5 meters in circumference. Imagine that the tires are rotating at 500 revolutions per minute.

1. What is the speed of the car?

2. Write verbal and symbolic rules that express the relation between time and distance.

3. Create a table that records some specific data pairs (time, distance) of your choice.

4. What would be the distance traveled in 20 minutes?

5. Sketch a graph that represents the data you have in your table in number 3. Graph the independent variable on the horizontal axis and the dependent variable on the vertical axis.

Daily Warm-Ups: Algebra, NCTM Standards

A Number Pattern?

Consider the mathematical sentences below. Are they true sentences? Is there a pattern? Will this pattern always work? If so, express the pattern in a general sentence with variables and show that it is a true sentence.

$$2^2 + 3^2 + 6^2 = 7^2$$
$$3^2 + 4^2 + 12^2 = 13^2$$
$$4^2 + 5^2 + 20^2 = 21^2$$

Daily Warm-Ups: Algebra, NCTM Standards

Thinking Recursively 1

Write recursive sentences that would generate each of the sequences below.

1. 1, 3, 9, 27, . . .

2. 2, 9, 16, 23, . . .

3. 36, 12, 4, $1.\overline{333}$, . . .

4. −12, −7, −2, 3, . . .

Thinking Recursively II

For each sequence below, write a recursive formula and then find the indicated term. Tell whether each sequence represents an arithmetic sequence, a geometric sequence, or neither.

1. Find the tenth term: 1, 5, 9, 13, . . .

2. Find the tenth term: 1, 1, 2, 3, 5, 8, 13, . . .

3. Find the seventh term: 100, 10, 1, 0.1, . . .

4. Find the eleventh term: 0, $\frac{1}{6}$, $\frac{1}{3}$, $\frac{1}{2}$, . . .

5. Find the fifteenth term: −2, 4, −8, 16, . . .

Daily Warm-Ups: Algebra, NCTM Standards

Thinking Recursively III

Write the first seven terms of each sequence, starting with t_1.

1. $t_1 = -18$, $t_n = t_{n-1} + 6$, $n \geq 2$

2. $t_1 = .5$, $t_n = 3t_{n-1}$, $n \geq 2$

3. $t_1 = 37.6$, $t_n = t_{n-1} - 4.2$, $n \geq 2$

4. $t_1 = 6$, $t_n = -\dfrac{1}{2} t_{n-1}$, $n \geq 2$

Daily Warm-Ups: Algebra, NCTM Standards

Parametric Equations

The equations below represent two parametric equations. Make a table of values for t, x, and y, and graph the curve represented by these parametric equations. Then write the pair of equations as a single equation expressing y in terms of x.

$$x = f(t) = t^2 - 4$$
$$y = g(t) = .5t$$

Daily Warm-Ups: Algebra, NCTM Standards

Analyzing the Quadratic Formula

Grace has found the following equation in her notebook:

$$x = \frac{-8 \pm \sqrt{8^2 - 4(1)(10)}}{2(1)}$$

She knows that she was using the quadratic formula to solve an equation and is wondering what the original equation might have looked like. Reconstruct her equation. Then find the solution (roots) of the equation. What are the *x*-intercepts for the graph of the equation?

Thinking About Parametric Equations

Think about parametric equations. What do they look like? What kind of situations might they model? Write a few sentences to explain your thinking.

Daily Warm-Ups: Algebra, NCTM Standards

Thinking About Parametric Trigonometric Equations

Think about the equation $\left(\dfrac{x}{5}\right)^2 + \left(\dfrac{y}{2}\right)^2 = 1$.

1. Describe the shape of the graph of this equation.

2. Write the equation in parametric form.

3. What if the graph was shifted 1 unit to the left and 2 units up? What would the nonparametric and parametric equations for this new graph look like?

Daily Warm-Ups: Algebra, NCTM Standards

Part 5: Understanding Patterns, Relations, and Functions

National Council of Teachers of Mathematics: "Instructional programs from pre-kindergarten through grade 12 should enable all students to understand patterns, relations and functions."

Expectations

Grades 6–8:
- Represent, analyze, and generalize a variety of patterns with tables, graphs, words, and, when possible, symbolic rules.
- Relate and compare different forms of representation for a relationship.
- Identify functions as linear or nonlinear and contrast their properties from tables, graphs, or equations.

Grades 9–12:
- Generalize patterns using explicitly defined and recursively defined functions.
- Understand relations and functions and select, convert flexibly among, and use various representations for them.
- Analyze functions of one variable by investigating rates of change, intercepts, zeros, asymptotes, and local and global behavior.
- Understand and perform transformations such as arithmetically combining, composing, and inverting commonly used functions, using technology to perform such operations on more-complicated symbolic expressions.

Part 5: Understanding Patterns, Relations, and Functions, *continued*

- Understand and compare the properties of classes of functions, including exponential, polynomial, rational, logarithmic, and periodic functions.
- Interpret representations of functions of two variables.

Polly's Popsicle-Stick Parallelograms

Polly is creating patterns with Popsicle sticks.

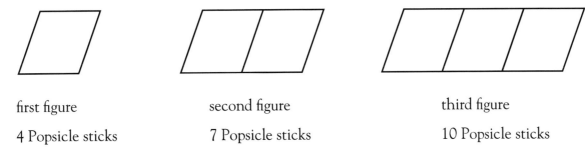

first figure

4 Popsicle sticks

second figure

7 Popsicle sticks

third figure

10 Popsicle sticks

1. How many Popsicle sticks are in the thirteenth figure in her pattern?

2. Write both an explicit equation and a recursive equation for the number of Popsicle sticks t_n for the *n*th figure in Polly's pattern.

3. What would a figure that has exactly 71 Popsicle sticks look like?

Daily Warm-Ups: Algebra, NCTM Standards

Triangular Numbers

Triangular numbers can be represented using dots, as shown below. The first triangular number is 1, the second is 3, the third is 6, and the fourth is 10, as pictured. How many dots will be in the fifth and sixth figures? Can you predict how many dots will be in the nth figure? Write an equation that could be used to determine the nth figure.

Patterns with Squares

Alexandra is babysitting her little brother. To entertain him, she is building structures with his blocks. The picture below represents three of the structures that she has built.

1. Sketch pictures of what you think the next two structures will look like if Alexandra continues the pattern that she has started. What number pattern is suggested by the number of blocks used in this sequence of structures?

2. How many blocks would be in the tenth structure if the pattern continues?

3. How many blocks would be in the 100th structure?

4. How many blocks would be in the nth structure?

Daily Warm-Ups: Algebra, NCTM Standards

Pentagonal Numbers

Pentagonal numbers can be represented using circles to form a pentagonal array as in the diagram below. Find a generalization for the nth pentagonal number.

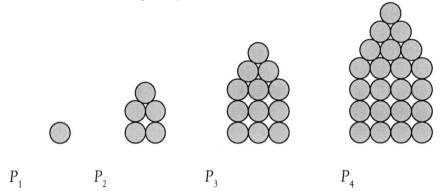

P_1 P_2 P_3 P_4

Daily Warm-Ups: Algebra, NCTM Standards

Newspaper Numbers

Many newspapers consist of large sheets of paper that are folded in half to form the pages. For example, a newspaper that is made from 2 large sheets of paper would have 8 pages. Consider the following questions about newspapers created in this way.

1. If there were 10 sheets of paper, how many pages would there be?

2. What page numbers would appear in the center, or on the innermost sheet of the newspaper? What is the sum of the numbers?

3. What is the sum of the page numbers that would appear on ANY one side of ANY sheet of the newspaper?

4. What is the sum of all the pages of the newspaper?

5. Write a rule in words or symbols that would give the sum of the pages for a newspaper with N sheets of paper.

Daily Warm-Ups: Algebra, NCTM Standards

© 2008 Walch Education

Stacking Cereal Boxes

Freya works part-time in her family's grocery store. Her dad has asked her to stack 45 cereal boxes in a display area for an upcoming sale. The boxes have to be stacked in a triangle with 1 fewer box in each row ending with 1 box at the top (like the sample stack shown below). How many boxes should Freya put in the bottom row? Suppose she had B boxes to stack in the display. Help her find a rule or formula to determine how many boxes she should put in the bottom row.

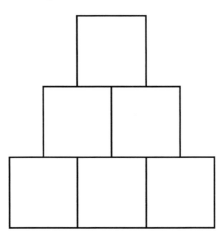

Pizza Cutting

Paolo loves pizza. He even dreams of pizza and makes up math problems about pizza. He wonders how many pieces of pizza he could get by cutting a very large circular pizza exactly 7 times. He decides that it doesn't matter if the pieces are the same size. He also decides that it isn't necessary for the cuts to go through the center of the pizza. How many pieces of pizza could he make? What if Paolo made n cuts in the pizza?

What is the greatest number of pieces he could make? Use the table below to help you.

Number of cuts	0	1	2	3	4	5	6	7	8	9		n
Number of pieces	1	2	4	7								

An Odd Pyramid

Look for patterns in the pyramid below. Then answer the following questions.

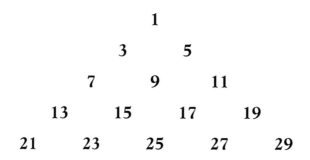

1. If the pyramid continues, what will be the middle number in row 25? In row 50? In row *n*?

2. What is the difference between the first and last number in row 4? In row 5? In row 40? In row *n*?

3. What is the last number in row 5? In row 6? In row *n*?

4. What is the sum of the numbers in row *n*?

5. What is the sum of all the numbers *through* row *n*?

Daily Warm-Ups: Algebra, NCTM Standards

How Long? How Far?

The Carmona family lives in Minnesota. They are driving to Florida for a vacation at an average speed of 60 miles per hour. Write an equation for a rule that can be used to calculate the distance they have traveled after any given number of hours. Then write a brief letter to the Carmona family that describes the advantages of having an equation, a table, and a graph to represent their situation.

Pool Border

Landscapers often use square tiles as borders for garden plots and pools. The drawing represents a square pool for goldfish surrounded by 1-foot square tiles.

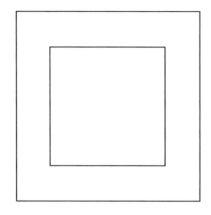

1. How many tiles will be needed for the border of this pool with an edge of length S feet?

2. Express the total tiles in as many different ways as you can. Be ready to explain why your different ways are equivalent.

Daily Warm-Ups: Algebra, NCTM Standards

Game Time

Abby and her brother Harry like to play a game called "U-Say, I-Say." Harry gives the "U-Say" number (an integer between −10 and +10). Abby has a secret rule she performs on the number that results in the "I-Say" number. Complete the table below by giving the missing "I-Say" values. Then describe Abby's rule in words and symbols.

U-Say	3	0	−4	1	2	5
I-Say	11	2	−10	5		

Daily Warm-Ups: Algebra, NCTM Standards

Number Puzzle

The difference between two numbers is 12. The product of the same two numbers is 45.

1. Find the two numbers using a symbolic strategy.

2. Find the two numbers using a graphing strategy.

3. Find the two numbers using a table.

4. Is there another way to find the two numbers?

5. Discuss the advantages and disadvantages of each strategy for this situation.

Daily Warm-Ups: Algebra, NCTM Standards

Mr. Wiley's Baby

Mr. Wiley presented some data to his algebra class about the early growth of his new baby boy. Hermione created a scatter plot for Mr. Wiley's data. The data and scatter plot are shown at right.

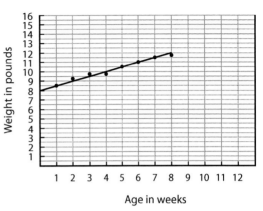

Week	Weight
1	8.5
2	9.25
3	9.75
4	9.75
5	10.5
6	11
7	11.5
8	11.75

1. Does the data seem to represent a linear model? What equation fits Hermione's graph model?

2. How long do you suppose this growth pattern will continue in this way? If it does continue, what would you expect Mr. Wiley's son to weigh in 1 year? In 2 years? In 14 years?

3. Within what age limits do you think this is a reasonable representation of the growth of Mr. Wiley's baby?

Daily Warm-Ups: Algebra, NCTM Standards

Sketch a Graph

Sketch a graph that represents each situation below. Identify your variables and label the axes of your graph accordingly.

1. Edgar is paddling his kayak down the reservoir to visit a friend. For half an hour, he paddled slowly, enjoying the morning. Then he noticed that the weather might be changing and increased his speed. After another 10 minutes, the wind picked up. For about 20 minutes, the wind blew in the direction he was paddling. Then it shifted and blew directly into his face for the last 15 minutes.

2. Jennifer has been taking tennis lessons for 3 months. She learned slowly at first, but her skill has been steadily increasing. Now she is going to visit her cousins in the country and won't be able to play or practice for two weeks.

Daily Warm-Ups: Algebra, NCTM Standards

Analyzing Scatter Plots

Ashley is reviewing the results of data collected and graphed from three different experiments. Help her by describing the patterns you see in these graphs. If appropriate, draw lines that you think might fit the data. If you don't think that a line fits the data, explain why.

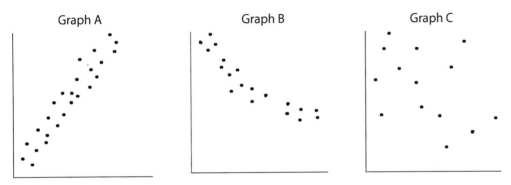

Daily Warm-Ups: Algebra, NCTM Standards

Identifying Graphs from Equations

Equations can reveal a lot of information. You can tell a lot just by how an equation looks and the variables and operations that are in it. Based on what you know about the nature of equations, identify whether each equation below represents a quadratic, a linear, or an exponential relationship. Explain how you made each decision.

1. $y = 7x + x^2$

2. $y = 3x + 7$

3. $y = (6 - x)x$

4. $y = 3^x$

5. $y = 3x(x + 4)$

Daily Warm-Ups: Algebra, NCTM Standards

Graphs to Sequences

Look at the graphs below. Identify each graph as a representation of either an arithmetic or geometric sequence. Use an informed guess to propose a possible recursive rule for each graph.

1. t_n

2. t_n

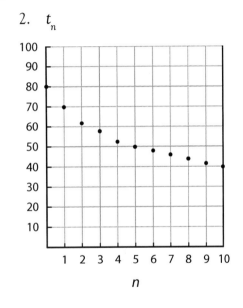

Daily Warm-Ups: Algebra, NCTM Standards

Explain Linear Regression

Write a brief but clear description of the conditions that would prompt you to pursue a linear regression equation on your graphing calculator. Explain your thinking. Describe what you might see on a graph of data points. Tell what keystrokes would be necessary to find the linear regression.

Forms of Quadratic Functions

John and Diana are studying quadratic functions. They have encountered the function below, written in three different forms:

$$f(x) = x^2 - 2x - 3$$
$$f(x) = (x - 3)(x + 1)$$
$$f(x) = (x - 1)^2 - 4$$

Show that the three equations are equivalent. How can you get each form from the other two? What different information is available to John and Diana from each of the equations? What are the zeros, or roots, of the equation? Without graphing and without using a calculator, give the coordinates of the vertex of the parabola and the x- and y-intercepts.

Daily Warm-Ups: Algebra, NCTM Standards

Graphing Basics

Serena is trying to remember some of the graphs for elementary functions. She is looking at the four graphs shown below. Help her by identifying a sample set of points on each graph. Then find the equation model that should represent each graph.

1.

2.

3.

4.

Table Patterns 1

Look at the tables below. For each table, do the following:

a. Describe symbolically or in words the pattern hiding in the table.

b. Give the missing values in the table.

c. Tell whether the relationship between x and y in the table represents a quadratic, a linear, an inverse, an exponential, or other relationship. Explain how you know.

1.

x	−1	0	1	2	3	4	5
y	5	7	9	11	13		

2.

x	−1	0	1	2	3	4	5
y	−12	0	10	18	24		

Daily Warm-Ups: Algebra, NCTM Standards

Table Patterns II

Look at the tables below. For each table, do the following:

a. Describe symbolically or in words the pattern hiding in the table.

b. Give the missing values in the table.

c. Tell whether the relationship between x and y in the table represents a quadratic, a linear, an inverse, an exponential, or other relationship. Explain how you know.

1.

x	0	1	2	3	4	5
y	1	3	9	27		

2.

x	−1	0	1	2	3	4	5
y	−5	0	3	4	3		

Daily Warm-Ups: Algebra, NCTM Standards

Table Patterns III

Look at the tables below. For each table, do the following:

a. Describe symbolically or in words the pattern hiding in the table.

b. Give the missing values in the table.

c. Tell whether the relationship between x and y in the table represents a quadratic, a linear, an inverse, an exponential, or other relationship. Explain how you know.

1.

x	−1	0	1	2	3	4	5
y		0	−4	−6	−6	−4	

2.

x	1	2	3	4	5
y	$1/2$	$1/4$	$1/6$		

Daily Warm-Ups: Algebra, NCTM Standards

Intercepting Intercepts

Do all quadratic functions have *x*-intercepts? Do they all have *y*-intercepts? Justify your answers to these questions by generating sample functions symbolically and/or graphically. Then explain if and why the functions have intercepts and what the graphs will look like.

Daily Warm-Ups: Algebra, NCTM Standards

Nonlinear Graph Models

Consider the equations and graphs below. The window is the same for each graph. Match each equation with the correct graph. Then explain why you made each choice.

1. $y = 8x^2$
2. $y = .1x^3$
3. $y = x^2$
4. $y = x^3$

a.

b.

c.

d.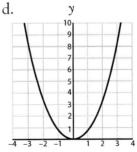

Daily Warm-Ups: Algebra, NCTM Standards

Thinking Absolutely

Think about each equation below. Without actually graphing (either with a calculator or by sketching), give the direction of opening, the vertex, and the equation that represents the angle of symmetry of the graph.

1. $y = |x| - 5$

2. $y = 3|x - 1| + 2$

3. $y = 4|x + 3| - 1$

Comparing the Graphs

1. Graph the functions $y = x^2$ and $y = \sqrt{x}$ together in the same convenient window. Then compare the graphs. How are they alike? How are they different?

2. What would the graph of $y = \pm\sqrt{x}$ be like? Would this graph still be a function? Why or why not?

Daily Warm-Ups: Algebra, NCTM Standards

Thinking About Quadratic Functions

Quadratic functions are often written in the form $y = ax^2 + bx + c$. Think about how changes in the values a, b, and c affect the graph of the function. Use the graph of $y = x^2$ to demonstrate the changes. Write a few sentences to explain your thinking.

Slides, Flips, and Stretches

Without actually graphing the sentence, explain how the graph model will change if the equation $y = x^2 - 1$ is changed in each way described.

1. y is replaced with $(y - 3)$

2. x is replaced with $-x$

3. x is replaced with $(x - 3)$

4. y is replaced with $\frac{y}{2}$

Daily Warm-Ups: Algebra, NCTM Standards

Graph Transformers

The graph to the right of $y = (x - 3)^2$ has been transformed into each graph below. Describe in words each transformation that has taken place. Then give a possible symbolic representation of the transformed graph. Each grid mark equals 1 unit.

1.

2.

3.

4.

Daily Warm-Ups: Algebra, NCTM Standards

Penny's Parabola

Penny has been given the equation $y = (x - 6)(x + 2)$ to graph. Penny's graph is shown below.

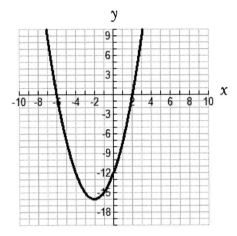

1. Using information about the x-intercepts, the vertex of the parabola, and other forms of the equation of the parabola, tell if you agree or disagree with Penny's graph.

2. Write Penny's equation in vertex form and in polynomial form.

Daily Warm-Ups: Algebra, NCTM Standards

A Quadratic Mystery

Without using a calculator to graph these equations, tell which ones are quadratic AND have a minimum point. Explain what you looked for and thought about as you reviewed the equations.

1. $y = x^2 + 5x + 6$

2. $y = x(9 - x)$

3. $y = (x + 3)(x - 2)$

4. $y = -9 - x^2$

5. $y = x(10 + 5)$

Daily Warm-Ups: Algebra, NCTM Standards

Digging for Roots

Find any roots or zeros of each quadratic function below. Then explain what roots or zeros mean graphically.

1. $f(x) = 2x^2 - 5x - 3$

2. $f(x) = x^2 + 2x + 1$

3. $f(x) = x^2 + 2x + 3$

4. $f(x) = 2x^2 + 3x - 1$

Daily Warm-Ups: Algebra, NCTM Standards

A Family of Functions

Consider the following three functions:

$f(x) = x^2 + 7x + 4$

$f(x) = -x^2 + 7x + 4$

$f(x) = ax^2 + bx + c$

What makes them a "family"? In other words, how are they related? What do they look like? Describe their rates of change, symmetry features, maximum or minimum values, and any other special features. What do the values a, b, and c tell you about the nature of the graph in this family?

Daily Warm-Ups: Algebra, NCTM Standards

Discriminating Discriminants

How much do you know about quadratic functions? Without graphing, discuss how many, if any, x-intercepts each quadratic function has.

1. $f(x) = x^2 + 3x + 4$

2. $f(x) = x^2 + 3x - 4$

3. $f(x) = 9x^2 - 6x + 1$

4. $f(x) = 3x^2 + 11x + 6$

Daily Warm-Ups: Algebra, NCTM Standards

Asymptote Adventure

Think about the functions given below. Determine the domain, range, and all asymptotes for each function's graph.

1. $f(x) = 7 + \dfrac{3}{x}$

2. $g(x) = \dfrac{1}{x-5} + 3$

3. $h(x) = \dfrac{2x + 3}{x}$

It's All in the (Rational) Family

Consider the following functions:

a. $r(x) = \dfrac{3}{x}$

b. $r(x) = \dfrac{3}{x^2}$

c. $r(x) = \dfrac{3}{x^3}$

d. $r(x) = \dfrac{3}{x^4}$

1. Predict what the graphs of these functions will look like without graphing them first. How will they be alike? How will they be different? What are their respective asymptotes? Is it appropriate to call them *functions*? Why or why not?

2. Now graph the functions using graph paper or a graphing calculator. Do you notice any additional characteristics?

Exponential Function Features

1. Consider the following functions:

   a. $g(x) = 3 \cdot 2^x - 1$

   b. $h(x) = 2 \cdot 3^x + 4$

   c. $k(x) = 2 \cdot \left(\dfrac{3}{2}\right)^x$

   Without graphing the functions, make some predictions based on your understanding of exponents. What happens to the graph of each function if x gets very large? What if x is negative? Where do these graphs cross the y-axis?

2. Now consider the function $f(x) = a \cdot b^x + c$. How do the values of a, b, and c affect the graph of the function? What if $a < 0$? What if $0 < b < 1$? What happens when you change b to $(1/b)$?

Intervals of Change

Look at the table and graph below.

x	y
−5	−10
−4	0
−2	13
0	16
2	14
4	2
5	−5

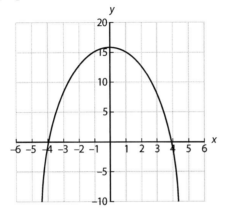

1. Identify the x- and y-intercepts, if any. How is the table helpful? How is the graph useful?

2. Consider the x interval (−5, 0). Describe how y may be changing over that interval of x.

3. Now consider the x intervals (0, 2), (2, 4), and (4, 5) for this graph and table. Is the average rate of change constant over these intervals?

Daily Warm-Ups: Algebra, NCTM Standards

© 2008 Walch Education

Functioning Properly?

Gwen and Rita are discussing the characteristics of certain functions. Rita claims that the equations $y = |x|$ and $y = x^2$ do not model functions because the line $y = 5$ intersects the graph of each equation in two points. Gwen disagrees. How might she present her case to Rita?

Another Functional Family

Consider the following three functions:

$h(x) = 3^x$

$h(x) = (0.3)^x$

$h(x) = a^x$

What makes them a "family"? In other words, how are they related? What do they look like? Describe their rates of change, symmetry features, maximum or minimum values, and any other special features. What features make this family different from other families? What effect do the constants have on the function and its graph?

The Step Function

Draw a graph of y = [x + 2] + 1. Why do you think this function is called a "step" function? What other name is applied to this type of function? Why?

Thinking About Holes and Asymptotes

Some equations create unusual results when they are graphed. Think about equations and graphs that might have a hole or an asymptote at $x = 2$. Give some examples. Then write a few sentences explaining how you can use an equation to predict where there would be a hole or an asymptote. How can you tell how to write an equation for a function from a graph with a hole or an asymptote?

Daily Warm-Ups: Algebra, NCTM Standards

Windows of Opportunity

Bob, Carol, and Ted are exploring the graph of $f(x) = \dfrac{2x^2 + 13x + 7}{x - 2}$. Bob believes that it represents a parabola. Carol thinks that it looks like a linear function. Ted doesn't know what to think. They have asked you to help them. The graphs they have drawn are shown below.

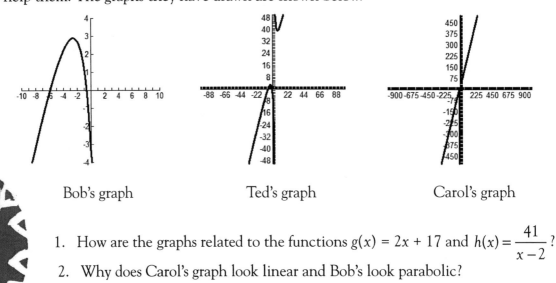

Bob's graph   Ted's graph   Carol's graph

1. How are the graphs related to the functions $g(x) = 2x + 17$ and $h(x) = \dfrac{41}{x - 2}$?
2. Why does Carol's graph look linear and Bob's look parabolic?
3. What happens to the graph of the function at $x = 2$?

Daily Warm-Ups: Algebra, NCTM Standards

Finding Inverses of Functions

What do we mean when we say that a function is an inverse of another function? Is the inverse of a function also a function? Write a few sentences about functions and their inverses. Then find the inverse function of each function below.

1. $f(x) = \dfrac{2}{5}(x+3)$

2. $g(x) = (x+2)^{2/3} - 6$

3. $h(x) = 6 - 2^x$

A Composition of Functions

1. Complete the table below for the following functions of x:

   $f(x) = x^2 + 1$

   $g(x) = (x-1)^2$

x	$f(x)$	$g(x)$	$f(g(x))$	$g(f(x))$
2				
		4		

2. Give the expressions that are equivalent to $f(g(x))$ and $g(f(x))$.

Daily Warm-Ups: Algebra, NCTM Standards

More Function Compositions

Use the following functions to find each value below:

$$f(x) = -3x + 5 \qquad g(x) = x^2 - 2 \qquad h(x) = (x - 1)^2$$

1. $f(3)$

2. $g(-2)$

3. $h(x + 1) - 3$

4. $f(g(3))$

5. $g(h(-1))$

6. $f(g(a))$

7. $h(f(a))$

8. $g(h(x))$

Daily Warm-Ups: Algebra, NCTM Standards

© 2008 Walch Education

Mystery Function

The graph below represents a function, F(x):

1. What is $F(-1)$?
2. What is $F(5)$?
3. $F(x) = 3$ is true for what values of x?
4. $F(x) > 4$ is true for what values of x?
5. What are the domain and range as represented by the graph of $F(x)$?

Daily Warm-Ups: Algebra, NCTM Standards

Analyzing Tables

Which tables of (x, y) values given below could represent functional relationships? Explain your thinking.

a.

x	y
0	5
1	7
2	8
3	10
4	5
5	7
6	8
7	9

b.

x	y
3	5
4	9
6	11
3	7
11	9
7	6
8	4
10	6

c.

x	y
4	11
8	7
7	3
9	5
5	12
3	11
2	8
6	9

d.

x	y
a	b
c	d
m	n
a	g
h	s
t	u
r	s
d	k

Daily Warm-Ups: Algebra, NCTM Standards

Exponential Function Mysteries

The following graphs were found without any units indicated. What can you say about the values of a and C for Graph 1? What can you say about the values of b and D for Graph 2? Explain your thinking.

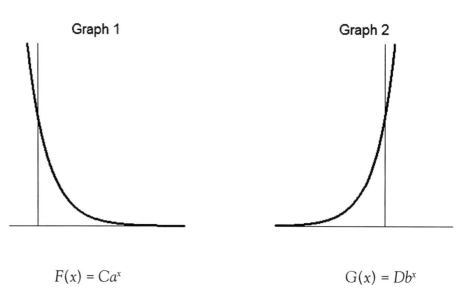

Graph 1: $F(x) = Ca^x$

Graph 2: $G(x) = Db^x$

Daily Warm-Ups: Algebra, NCTM Standards

Mystery Table

Sophia and Tran have discovered this partially completed table in their algebra notes. Sophia thinks it represents a quadratic function. Tran thinks it represents exponential decay. Do you agree with either of them? If so, who and why? If not, why not? Does the table represent some other kind of function? Complete the table and write a symbolic sentence for the function that you think is represented by the table values.

x	y
−2	
−1	
0	64
1	12.8
2	2.56
3	.512
4	

Daily Warm-Ups: Algebra, NCTM Standards

Absolute Graphing

Draw the graph of $|x| + |y| = 5$. Then consider each change to the original equation as shown below. Make a prediction of how the graph will change. Then verify your prediction by graphing.

a. $|x| - |y| = 5$

b. $|y| - |x| = 5$

c. $|x + y| = 5$

Graphing Trigonometric Functions

The graph of $y = \cos(x)$ is shown below. Use your knowledge of transformation of functions to sketch the graph of one cycle of each equation that follows.

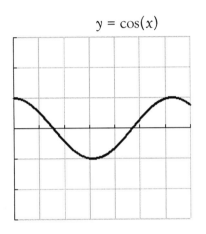

$y = \cos(x)$

1. $y = 2 + \cos(x)$
2. $y = \cos(x - \pi)$
3. $y = 2\cos(x - \pi) - 1$

Daily Warm-Ups: Algebra, NCTM Standards

Looking for Change

In the following sketch, the solid black curve fits the function $f(x) = x^4 + x^3 - 3x^2$. Using this information, determine the function sentence for the dashed-line graph and the dotted-line graph. Justify your thinking and check your answer using a graphing calculator.

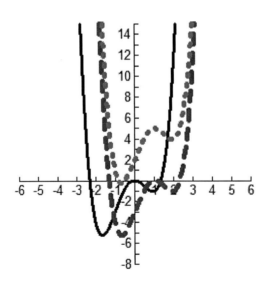

Daily Warm-Ups: Algebra, NCTM Standards

Answer Key

Part 1: Developing Algebraic Thinking Through and for Mathematical Modeling

1. There is a difference of 2 between the two products. The numbers can be represented as $n, n + 1, n + 2, n + 3$; $n(n + 3) = n^2 + 3n$ represents the outer products. $(n+1)(n+2) = n^2 + 3n + 2$ represents the inner products, thus the inner product will always be 2 more than the outer product.

2.
Number of purple rods	1	2	3	4	5	6	n
Surface area	18	28	38	48	58	68	$10n + 4$

3.
Number of purple rods	1	2	3	4	5	6	n
Surface area	18	30	42	54	66	78	$18 + 12n + 2$

4. Answers will vary. Sample answer:

S	Area of square	R	Area of circle
1	1	1	3.141592654
2	4	2	12.56637061
3	9	3	28.27433388
4	16	4	50.26548246
5	25	5	78.53981634
6	36	6	113.0973355
7	49	7	153.93804
8	64	8	201.0619298
9	81	9	254.4690049
10	100	10	314.1592654

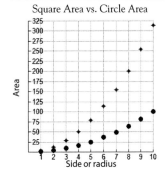

Daily Warm-Ups: Algebra, NCTM Standards

Answer Key

5. Students may need to be reminded to convert 12.5 feet to 381 centimeters. Students' equations may vary depending on the strategy used. This data and graph leads to an answer of 23 rubber bands.

6. H = 120

Number of students (n)	Number of handshakes (h)
1	0
2	1
3	3
4	6
5	10

$H = \dfrac{n(n-1)}{2}$ where n = number of people

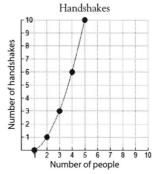

7. 33 people provide the greatest income for EVA.

Number of people	10	14	18	22	26	30	31
Cost per person	200	184	168	152	136	120	120
Total income	2000	2576	3024	3344	3536	3600	3720

Number of people	32	33	34	35	36	37	38
Cost per person	120	120	104	104	104	104	88
Total income	3840	3960	3536	3640	3744	3848	3344

Daily Warm-Ups: Algebra, NCTM Standards

Answer Key

8. 1.

   2. Linear regression $(a + bx)$
      regEQ$(x) = -20.3 + 15.3x$
   3. Using a linear model as a predictor indicates that Linn might have 209 mice after 15 months. Using an exponential graph as a predictor indicates that there may be 200 mice by 7 months.

9. linear regression $(a + bx)$
   regEQ$(x) = 59.0067 + -.003567x$

10. 1. $A = (n(120 - 2n))$ expresses the area of the pen when one side is against the barn, where n = width. ($n = 30$; $m = 60$)
    2. $n = 40$; $m = 40$; $A = 1,600$
    3. The maximum area using the barn is 1,800 square feet.
    4. Minimum values will vary depending on students' decisions on how short a width can be. One possible answer is 118 square feet using the barn.

Daily Warm-Ups: Algebra, NCTM Standards

5. The maximum area without using the barn is 900 square feet. Minimum areas will vary; one possible answer is 59 square feet.

11.

Side of square cutout(s)	Length of box $(12 - 2s)$	Width of box $(9 - 2s)$	Volume of box $s(12 - 2s)(9 - 2s)$	Surface area of box
1	10	7	70 in^3	104 in^2
2	8	5	80 in^3	92 in^2
3	6	3	54 in^3	72 in^2
4	4	1	16 in^3	44 in^2

Greatest volume results from 2-inch cutouts. Smallest volume is from 4-inch cutouts.

12. 1. $x + y \leq 75$ and $y \leq .5x$ and $y \geq 0$ and $x \geq 0$
2.

3. The intersection of the shading identifies the distribution of fiction and nonfiction books for her purchases.
4. Marian can buy 50 nonfiction and 25 fiction books.

Daily Warm-Ups: Algebra, NCTM Standards

Answer Key

13.

Toppings and soft drink

Corner points: (0, 0), (0, 16) and (32, 0). Boundaries: $.5t + 1.00s \leq 16$ and $t \geq 0$, and $s \geq 0$

14. A table and graph might look like this:

Day	Salary	Uncle's commission
1	1,000	100
2	1,800	200
3	3,200	400
4	5,600	800
5	9,600	1,600
6	16,000	3,200
7	25,600	6,400
8	38,400	12,800
9	51,200	25,600
10	51,200	51,200
11	0	102,400
12	–102,400	204,800
13	–204,800	409,600

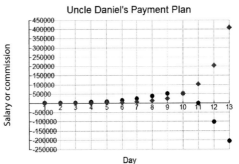

Uncle Daniel's Payment Plan

Daily Warm-Ups: Algebra, NCTM Standards

It might not be wise to continue to work for Uncle Daniel after day 10.

15. The unit cube has 8 corners, 6 faces, and 12 edges.
    $T = n^3$

L1	L2
1	1
2	8
3	27
4	64
5	125
6	256
7	343
8	512
9	729
10	1000

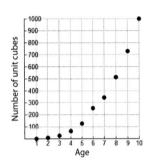

16. 1.

Side length	3 faces	2 faces	1 face	Zero faces
2	8	0	0	0
3	8	12	6	1
4	8	24	24	8
5	8	36	54	27
6	8	48	96	64
7	8	60	150	125
8	8	72	216	216
9	8	84	294	343
10	8	96	384	512
15	8	156	1014	2197

    2. 3 faces located in corners; 2 faces located on edges; 1 face located on faces other than those listed above; 0 faces located on the interior
    3. For all cubes, 3 faces is 8; 2 faces: $N = 12(s - 2)$; 1 face: $N = 6(s - 2)^2$; zero faces: $N = (s - 2)^3$.

17. 1. $2,030.00
    2. $2,060.45
    3. $2,090.90
    4. After 15 months, his unpaid balance will be greater than $2,500.

18. 1. inverse; $xy = 5$
    2. inverse; $xy = 12$
    3. direct; $y = 2x/3$

Daily Warm-Ups: Algebra, NCTM Standards

Answer Key

19.
   4. Neither; there is no constant of proportionality.
   1. a. 4; b. 6; c. 12; d. –3; e. 0.5
   2. Answers will vary.

20. Answers will vary. For example, students may refer to symmetry with respect to the y-axis. They may relate even powers of n in the function $y = x^n$ to $y = x^2$ and odd powers to $y = x^3$. They may talk about the graph opening upward or downward, about the general slope of the function, and about the growth of values in a table of these functions.

21. Answers will vary. Sample answer: For $a > 0$, would appear in quadrants I and III; for $a < 0$, it would appear in quadrants II and IV. For $a > 0$, $y = \frac{a}{x^2}$ would appear in quadrants I and II; if $a < 0$, it would appear in quadrants III and IV. Students might also discuss asymptotes and where the functions are increasing and decreasing.

22.
   1. 23 toothpicks
   2. explicit equation: $t_n = 2n + 1$; recursive equation: for $t_1 = 3$, $t_n = t_{n-1} + 2$

23.
   1. 15 moves
   2.
Number of each color	1	2	3	4	5	6	7	8	9	10
Number of moves	3	8	15	24	35	48	63	80	99	120

   3. $M = n(n + 2)$

   3. There would be 35 connected triangles.

24.
   1. In the seventh week, the store will be selling at a loss.
   2. Recursive formula: t_1 = first term; $(tn) = t_{n-1}(r)$, $n > 1$

Term number (n)	Term (t)	Recursive form of t_n
1	30	30
2	27	30 × .90
3	24.30	27 × .90
4	21.87	24.30 × .90
5	19.68	21.87 × .90
6	17.71	19.68 × .90
7	15.94	17.71 × .90

   3. Yes, the t_n values in the table above represent a geometric sequence. Each value is found by multiplying the preceding term by a constant value. In this case, the constant value is .90 (or 1.00 – 10).

Daily Warm-Ups: Algebra, NCTM Standards

25. Juan is correct. These situations can be represented as inverse functions.
Tables:

l	w
36	1
18	2
12	3
9	4
6	6
4	9
3	12
2	18
1	36

l	w
24	1
12	2
8	3
6	4
4	6
3	8
2	12
1	24

Sample graph:

Possible equations: $lw = 24$, thus $y = \frac{24}{x}$ could be a representation of this function for $x = w$ and $0 < w \leq 24$.

26. 1. $3.10
    2. $2.86
    3. $y = 2.98(1+.04)^x$
    4. In year 5, it will surpass $3.50.

27. $I = (.20 + .02x)(10000 - 200x)$, where x is the number weeks to wait; $x = 20$ weeks.

28. Car B will be worth more than Car A after 3 years.

Year	Car A	Car B
0	15,000	12,000
1	15,000 × .70 = 10,500	12,000 × .80 = 9,600
2	7,350	7,680
3	5,145	6,144

29. 1. $C(d) = 1,200(2^d)$, Count $= 1,200(2^d) = 1,200(2^{3.25})$
       $= 11,416$
    2. A trace can be used to find the count at $x = 3.25$ for the number of days indicated in the question and for when the bacteria count goes over 200,000.
       $200,000 = 1,200(2^d)$
       $2^d = 166.66$
       $d \approx 7.4$
       Thus, the pool will need treatment on day 7.

Daily Warm-Ups: Algebra, NCTM Standards

Answer Key

30. $B(x) = 200(1-.25)^x$. Bounce 5 will be less than 50 centimeters.

Bounce	Height
0	200
1	150
2	112.5
3	84.375
4	63.281
5	47.46
6	35.6
7	26.7
8	20
9	15
10	11
11	8

31. $76 < (25 + w)(50 + w) - 25(50) \leq 400$; Solving the two resulting inequalities results in lengths for w that could be between 1 meter and 5 meters.

32. $C = 20{,}000 + .02p$, $C = 17500 + .025p$; The school must make more than 500,000 copies before the higher price is a reasonable choice.

33. Sidney's salary is a geometric progression in which $t_1 = 35{,}000$, $r = 1.05$, $t_n = 1.05 t_{n-1}$, and $1 \leq n \leq 4$.

Daily Warm-Ups: Algebra, NCTM Standards

Answer Key

Sidney's rent is an arithmetic progression in which $t_1 = 950$, $d = 50$, $t_n = t_{n-1} + 50$, and $1 \leq n \leq 4$. Explanations will vary.

34. $A = LW$ and $L + 2W = 1200$, $\dfrac{L + 2W}{2} \geq \sqrt{2LW}$, $600 \geq \sqrt{2A}$, $A \leq 180{,}000$

35. Solutions will vary depending on students' choices. One graph and linear regression equation are provided for the data for both sexes.
    Linear regression ($ax + b$)
    regEQ(x) = .229738x + −381.817
    a = .229738
    b = −381.817
    r = .981234
    r^2 = .96282
    For this graph and equation, a student might predict that a person born in 2020 would have a life expectancy of 82.26 or 82.3 years.

36. 1. b
    2. c

37. 1. c
    2. a

38. Students should choose to justify choice c as the proper graph.

39. 1. between 4 and 6 seconds
    2. about 13 mps; 48 mps
    3. from 1 to 4 seconds into the ride; 6 to 7 seconds into the ride
    4. The velocity increased rapidly. They stayed at 55 mps for 1 second, then decreased between 7 and 8 seconds, increased to about 53 mps, and then began to decrease again.

40. 1. 2.03 million per year from 1900 to 2000 is less than the 2.8 million per year from 1990 to 2000.
    2. Highest rate of change: 1950 to 1960 or 1990 to 2000. Lowest rate of change: 1930 to 1940. Calculate slope to find rates of change. Factors might include the postwar baby boom, the Great Depression, and so forth.
    3. Answers will vary. Sample answer: 325,000,000

Daily Warm-Ups: Algebra, NCTM Standards

Answer Key

41. 1. 115° C
    2. 36° C
    3. 20° C
42. $C_1 = 20.00 + 10m$; $C_2 = .40m$

Minutes	Horizon	Best Talk
0	0	20
10	4	21
20	8	22
30	12	23
40	16	24
50	20	25
60	24	26
70	28	27
80	32	28
90	36	29

In the graph, Best Talk is represented by the diamond shapes and Horizon is represented by the circles. Students may argue that Horizon is the best plan until Aisha exceeds 65 minutes in a month.

43.
Original square		Rectangular plot			Difference in areas
Side length	Area	Length	Width	Area	
4	16	7	1	7	9
5	25	8	2	16	9
6	36	9	3	27	9
7	49	10	4	40	9
8	64	11	5	55	9

This is not a fair trade with respect to area. The rectangle will always be 9 square meters smaller than the square. The area of a square = n^2. The area of a rectangle = $(n+3)(n-3) = n^2 - 9$.

44. 1. $T = 2000 + 399m$; $T = 1600 + 420m$
    2. At the end of 1 year, Joe's car will have cost $6,788 and Hank's car will have cost $6,640. At the end of 2 years, Joe's car will have cost $11,576, and Hank's car will have cost $11,680.
    3. Answers will vary. Students could make an argument for either plan.

Daily Warm-Ups: Algebra, NCTM Standards

Answer Key

45. 1.

2. area of square = s^2, area of rectangle = $(s+5)(s-4)$

s	Area of square	Area of rectangle
5	25	10
10	100	90
15	225	220
20	400	400
25	625	630
30	900	910
35	1,225	1,240

3. For $s > 20$, the area of the rectangle will be greater than the area of the square.
4. For $s < 20$, the area of the rectangle will be less than the area of the square.
5. For $s = 20$, the areas will be equal.
6. Answers will vary.

46. Substitution into the given function or graphing and tracing yields results to the question. Graphing: $H(t) = -16t^2 + 120t + 6$; the highest point is 231 feet after 3.75 seconds. The ball will stay in the air for approximately 7.5 seconds.

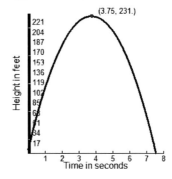

47. 1. $1,113.94
2. more than 370 months (or 30 years, 10 months)
3. 185 months (or 15 years, 5 months)

Daily Warm-Ups: Algebra, NCTM Standards

Answer Key

48.

x	y
0	200
1	164
2	134.48
3	110.27
4	90.42
5	74.15
6	60.80

The 200 represents the measure of the starting quantity, and the 0.18 represents an 18% decay rate.

49. The time is 2.02 seconds when the height is 10 meters. Students may use a graphing calculator and the trace feature to find the answer.

50. 1. Marcus started with 800 bacteria.
    2. This window shows $0 \le t \le 30$ and $0 \le p \le 800$. Students may make other reasonable choices. The graph represents a function. Any vertical line will intersect the graph in one and only one point.

   3. Half-life can be found in one of three ways. Students might trace to find the t value closest to 400 as shown on the graph. They could also find the intersection of this function with the line $y = 400$. They might use a table of values to find the time when the bacteria population is nearest 400.

Daily Warm-Ups: Algebra, NCTM Standards

Answer Key

51. 1. $h(3) = -4.9(3)^2 + 150 = 105.9m$. This represents the height of the tennis ball 3 seconds after it is released.
    2. $-4.9t^2 + 150 \leq 25$, $t \geq 5.05$ sec
    3. $(-4.9t^2 + 150 = 0, t); t = 5.53283, t = -5.53283$. In this problem, -5.53 has no meaning.

52. Students could solve the equation substituting 12 for h, or they might draw a graph on their graphing calculators and estimate the intersection of two lines. For example, the graph below indicates that the height will be 12 feet at approximately 7 A.M. and again at 11 A.M. Melinda's father is too late by 1 hour in his prediction. The height will only be about 9.5 feet.

Part 2: Analyzing Change in Various Contexts

53. 1. slope = $-600/2$ or -300
    2. rate of change = 300 mps downward
    3. The domain for the line segment is $1 \leq s \leq 3$. The range for the line segment is $1800 \leq A \leq 2400$.

54. 1. $C = 4.75(n) + 1.00$, C = total cost, n = number of packages
    2.
    3. The shipping price increased by $0.50.
    4. $C = 4.75(n) + 1.50$

Daily Warm-Ups: Algebra, NCTM Standards

Answer Key

55. Graphs will vary depending on students' interpretations of the situations. Most students will create line or bar graphs. Discuss the appropriateness of a continuous or discrete graph for these models.

56. Answers will vary. A graph of the combined data is given. An argument could be made for either plan. If Rosa's family rents 25 DVDs per month, but they only watch 2 movies per month on average, the Deluxe company is a better choice for her family.

57. 1.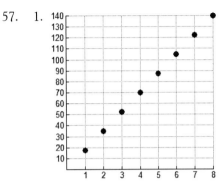

    It would not make sense to connect the points on the graph. The data points are discrete, not continuous.

    2. As the number of campsites increases, the fee increases a constant amount. The graph shows this with a slope up and to the right.

Daily Warm-Ups: Algebra, NCTM Standards

Answer Key

58.

Number	Total Charges
10	190
15	285
20	380
25	475
30	570
35	665
40	760
45	855
50	950

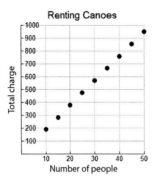

59.
1. Variables are time and number of step-ups.
2. As time increases, the number of step-ups in each 10-second interval decreases.

y	# step ups	11	19	29	39	48	52	59	62	65	69	72	75
x	time in seconds	10	20	30	40	50	60	70	80	90	100	110	120

3. Estimates: (25, 25), (65, 55); explanations will vary.

60.
1. A table of values would have increasing values of x and simultaneously increasing values of y. The graph would slope up and to the right.
2. This situation would have x values increasing and y values decreasing, so the slope of the graphed line would be down and to the right.
3. The points lie on a straight line when there is a common slope between each set of points.
4. The points could be connected if the data expressed represents a continuous change in values.

61. The weight increases more each month in the first year than in the second year. The graph looks like it might fit a parabolic curve for $0 \leq x \leq 25$.

Daily Warm-Ups: Algebra, NCTM Standards

Answer Key

Quadratic regression
regEQ(x) = −.035955x^2 + 1.63514x + 7.82424

62. Answers will vary. Sample answer: Both cars are traveling the same speed at $T = 1$ hour, but the car represented by the dotted line will be some distance ahead of the other car. However, the car represented by the solid line is accelerating faster than the other at $T = 1$ hour.

63. a. The person started approximately 2.5 units (feet or meters) from the detector, walked away from the detector rapidly at first and then more slowly, and stopped when he or she was approximately 5.5 units away from the detector.
b. The person started 2 units away from the detector and walked away at a constant rate of 1/2 unit per second.
c. The person started 1 2/3 units away and walked away slowly at first and then more rapidly. The person stopped approximately 5 2/3 units away from the detector.

Daily Warm-Ups: Algebra, NCTM Standards

Answer Key

64. 1.

2.

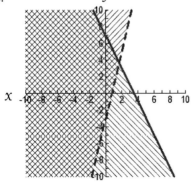

65. 1. $3x + y \geq 6$ and $3x + y \leq 10$
    2. $x \geq -3$ and $y \leq 5x/2 + 5$ and $y \leq -5x/3 + 5$
66. 1. arithmetic, linear, decreasing
    2. geometric, nonlinear, decreasing
    3. arithmetic, linear, increasing
    4. geometric, nonlinear, decreasing
    5. geometric, nonlinear, increasing
67. At 80 miles per hour, the stopping distance would be close to 500 feet. Explanations will vary.

Daily Warm-Ups: Algebra, NCTM Standards

Answer Key

Part 3: Using Algebraic Symbols I

68.

Number of toothpicks	6	11	16
Number of hexagons	1	2	3
Perimeter of figure	6	10	14

$T = 5H + 1$; $P = 4H + 2$

69.
1. -3
2. $-5\frac{1}{2}$
3. 128
4. 4.75
5. 0
6. 72
7. 1
8. 16
9. 2
10. 50

70.
1. 24
2. 10
3. 413
4. 24
5. 12.5
6. -90
7. 3/4
8. 0
9. 0
10. -40

71.
1. $S = 4L + 4W + 4H = 4(L + W + H)$
2. $A = 2LW + 2LH + 2WH = 2(LW + LH + WH)$

72.
1. $2(5) + 2(4) + 4(2.5) = 28$ meters
2. $P = 2M + 2N + 4L = 2(M + N) + 4L$
3. Answers will vary.

73. Students might talk about undoing the steps and subtracting 17 from (or adding -17 to) both sides and then dividing by -4.

74.
1. 4
2. 6
3. 1
4. 0, 3.5
5. 0, 13
6. 0, -2
7. 0, 5
8. -2
9. -2
10. 0, 7

Daily Warm-Ups: Algebra, NCTM Standards

Answer Key

75.
1. $8x - 40$ or $4(2x - 10)$
2. $5x^2 - 6x + 13x - 10 = 5x^2 + 7x - 10$
3. $4x + 20 - 6 + 12x = 16x + 14$
4. $30 - 5x + 5x + 5 = 35$
5. $-5x^2 + 3x + 5$ or $5(1-x^2) + 3x$

76.
1. $y = 8x - 21$; slope = 8, y-intercept = $(0, 21)$
2. $y = x - 1$; slope = 1, y-intercept = $(0, -1)$
3. $y = \frac{5}{3}x + 6$; slope = 5/3, y-intercept = $(0, 6)$
4. $y = 2.5x + 13$; slope = 5/2, y-intercept = $(0, 13)$
5. $y = 13.2x - 138.4$; slope = 66/5, y-intercept = $(0, -138.4)$

77.
1. $y = -3x + 11$
2. $y = -\frac{2}{3}x + \frac{5}{3}$
3. $y = -\frac{5}{4}x + \frac{17}{2}$
4. $y = -x + 1$
5. $y = x + 0$

78.
1. $y = \frac{3}{2}x + 2$
2. $y = 3$
3. The line passes through $(1, 5)$ and $(4, 6)$, thus $y = \frac{1}{3}x + \frac{14}{3}$.
4. $y = -x$

79.
1. $y = 2x + 3$
2. $y = -3x + 5$
3. $(y - 5) = \frac{1}{3}(x - 2)$
4. $(y - 7) = \frac{2}{3}(x - 3)$

80.
1. $y = -300x + 350$ for x = cost and y = number sold
2. The slope = -300, and the y-intercept = 350. For every $1.00 increase in price, the number sold decreases by 300; in addition, they can give away 350 brownies if they don't charge anything.
3. 140 brownies
4. about $0.17 for each brownie

81. 1–3. Answers will vary.

82.
1. a line of best fit through the points
2. a constant rate of change for the ordered pairs
3. first-degree values for x and y
4. Answers will vary.

Daily Warm-Ups: Algebra, NCTM Standards

Answer Key

83. Answers will vary. Sample answer: If you know the slope and y-intercept, you can substitute the values into the equation $y = mx + b$ or $y = a + bx$. The slope can be determined from the graph by finding the change in y divided by the change in x. The y-intercept will be the point where $x = 0$ in the table and where the line crosses the y-axis on the graph.

84. Answers will vary. Sample answer: The rate of change is the value found when the difference in two y values is divided by the difference in the corresponding two x values, either from the table or from the graph using any two points. In the equation, the rate of change is the value m or b when the equation is written in the form $y = mx + b$ or $y = a + bx$. The y-intercept is the value $(0, x)$.

85.
    1. $x + 2$
    2. not possible
    3. b
    4. $\frac{1}{f}$
    5. $x + 5$
    6. $n - 4$

86. Answers will vary. Sample answer: One way to compare two linear equation models is to compare their average mean errors by calculating the difference between the predicted value and the actual value for various x values.

87.
    1. $C = f(T) = T - 37$
    2.

Cricket Chirps

    3. Answers will vary.

88. Slope is $\frac{212-32}{100-0} = \frac{180}{100} = \frac{9}{5}$, thus $y = \frac{9}{5}x + 32$. Note: y = Fahrenheit and x = Celsius.

Daily Warm-Ups: Algebra, NCTM Standards

Answer Key

89. 1. Eric weighs about 104 pounds.
    2. If Amber and Aleah continue to sit at 4 feet, then $130 \times 4 = 104 \times D$, and $D = 5$ feet. If the seesaw is long enough, then Eric can balance them. Otherwise, he cannot.

90.

Number of subscriptions sold	Weekly pay (dollars)
1	145
2	170
3	195
6	270
10	370
12	420
15	495

   2. $W = 25S - 120$; $S = (W - 120)/25$
   3. 20 subscriptions

91. Answers will vary. Sample answers:
    1. C might represent the original cost of a coat with a discounted price of $37.49. $C = \$49.99$.
    2. The sale price of a shirt might be $28.00 after a 20% discount. C represents the original cost of the shirt. $C = \$35.00$

92. $70m + 100 > 500$; $70m > 400$; $m > 5.7$. School started 6 months ago.

Part 4: Using Algebraic Symbols II

93. Maria may be counting all the tiles at the top and bottom of the border $(n + 2)$ and then adding the remaining tiles on each side (n) for her expression. Carlos may be thinking that the border can be separated into 4 equal sets of $(n + 1)$ sections.

94. 1. no
    2. yes
    3. no
    4. yes
    5. no

    Circumference of circle, c, is 2, but Rajah is given a semicircle. Therefore, the circumference of the semicirle is πr. The rectangle has 3 sides to consider for the perimeter. The length is $2r$ and the height is r. Therefore, the perimeter of the rectangle $2h + 1 = 2r + 2r = 4r$. The total perimeter of the figure is $4r + \pi r$, or $r(4 + \pi)$.

Daily Warm-Ups: Algebra, NCTM Standards

Answer Key

95. **b and c**
    The volume of a cylinder is given by the formula $v = \pi r^2 h$ where πr^2 = base area. The height, h, of the cylinder can be represented by $x + y$. The volume is then expressed as $v = 45(x + y) = 45x + 45y$.

96. 1. $\frac{1}{100}$
    2. $\frac{99}{100}$
    Explanations will vary.

97. 1. Molly might be thinking about finding the break-even prices to charge. Lucas might be thinking about the fixed cost and about the change in sales if the ticket price is increased. Yes, the equations are equivalent.
    2. Break-even prices are $4 and $16. Fixed cost is $4,800. For every $1 increase in ticket price, 75 fewer tickets would be sold. The best price might be $10. Explanations will vary. Some students might use the trace feature on their calculator to determine the maximum values.

98. 1. $\begin{bmatrix} 1 & 10 \\ 9 & 1 \end{bmatrix}$
    2. $\begin{bmatrix} -8 & 16 \\ 20 & -4 \end{bmatrix}$
    3. not possible
    4. $\begin{bmatrix} 24 & -12 \\ -15 & 21 \end{bmatrix}$
    5. $\begin{bmatrix} -1 & -5 \\ 10 & -24 \end{bmatrix}$

Daily Warm-Ups: Algebra, NCTM Standards

Answer Key

99. 1. $\begin{bmatrix} 6 & 36 \\ 31 & 5 \end{bmatrix}$

   2. $\begin{bmatrix} -16 & 14 \end{bmatrix}$

   3. not possible

   4. $\begin{bmatrix} 2 & -23 & 11 \\ 13 & -26 & 24 \\ 3 & 11 & -1 \end{bmatrix}$

100. Answers will vary. Sample answer: An advantage is that matrices are easily stored and calculated with technology. Disadvantages are that matrices only represent two variables in the rows and columns and are not as visual as graphs. A matrix could be represented in different matrices by transposing the rows and columns and by listing the data in a different order.

101. Answers will vary. Students should mention the importance of the dimension conditions. Yes, order matters: $A \times B \neq B \times A$.

102. Answers will vary. Sample answer: Matrices can be added if they have the same dimensions. They can be multiplied if the columns in matrix A match the rows in matrix B. The dimensions of the product matrix will be the number of rows in A × the number of columns in B. If the determinant of the matrix is 0, then it has no inverse.

   $A = \begin{bmatrix} 2 & 4 \\ 3 & 6 \end{bmatrix}$

   The inverse of matrix $A = \begin{bmatrix} 5 & 3 \\ 4 & 2 \end{bmatrix}$ is $A^{-1} = \begin{bmatrix} -1 & \frac{3}{2} \\ 2 & -\frac{5}{2} \end{bmatrix}$.

   The opposite of matrix A is $\begin{bmatrix} -5 & -3 \\ -4 & -2 \end{bmatrix}$. The identity matrix for these cases is $\begin{bmatrix} 1 & 0 \\ 0 & 1 \end{bmatrix}$.

Daily Warm-Ups: Algebra, NCTM Standards

Answer Key

103.
1. $2\sqrt{3}$
2. $\dfrac{\sqrt{6}}{3}$
3. $\sqrt{5}$
4. $11\sqrt{2}$
5. $15\sqrt{3}$
6. $\dfrac{-13i}{13} = -i$
7. $\dfrac{(a+bi)^2}{a^2+b^2}$

104.
1. $\dfrac{1}{x^n}$
2. y^{r-s}
3. x^{ab}
4. $a^n b^n$
5. $\dfrac{d^n}{t^n}$
6. x^{-n}
7. n^{p+q}
8. $\sqrt[n]{y}$

105.
1. $x + y = 10$; $x + 2y = 8$; $(12, -2)$
2. $x + 2y$; $x + y = 15$; $(10, 5)$
3. $x - y = 2$; $2x - 2y = 4$; Any values for x and y that differ by 2 will work in this puzzle. There are infinite solutions.

106. Word puzzles will vary.
1. $(2, 1)$
2. $(-2, 4)$
3. not possible

107. Students' choices and rationales may vary.
1. $(-4, -5)$
2. $(-6, 4)$
3. $(-20, 10)$
4. $(3.6, 2.4)$

108. Answers will vary.
1. Sample answer: to find an ordered pair that satisfies both equations

Daily Warm-Ups: Algebra, NCTM Standards

Answer Key

2. Sample answer: Graph both lines on the same set of axes, and locate the intersection point of the two lines.

3. Sample answer: Substitute values of x into both equations. Look for identical y values for the same value of x.

4. Sample answer: solve algebraically by substitution or elimination

5. Sample answer: If the values in the table have the same constant rate of change, if the lines are parallel, or if the slopes are the same but have different y-intercepts, then there is no solution.

109.
1. $y = -16t(t + 3)$
2. $y = 5x(x - 2)$
3. $y = (3 + x)(x - 1)$
4. $y = 13(4x^2 - 1) = 13(2x - 1)(2x + 1)$
5. $y = (x + 4)(x + 1)$
6. $y = 2x^2 + 5x + x^2 - 10x = 3x^2 - 5x = x(3x - 5)$

110. $\begin{bmatrix} 6 & 4 \\ 1 & 1 \end{bmatrix} \begin{bmatrix} b \\ s \end{bmatrix} = \begin{bmatrix} 1750 \\ 400 \end{bmatrix}$;

$\begin{bmatrix} b \\ s \end{bmatrix} = \begin{bmatrix} 6 & 4 \\ 1 & 1 \end{bmatrix}^{-1} \begin{bmatrix} 1750 \\ 400 \end{bmatrix} = \begin{bmatrix} \frac{1}{2} & -2 \\ -\frac{1}{2} & 3 \end{bmatrix} \begin{bmatrix} 1750 \\ 400 \end{bmatrix} = \begin{bmatrix} 75 \\ 325 \end{bmatrix}$;

$b = 75, s = 325$

111.
1. correct
2. $5x$
3. $6n^2$
4. $6xy$
5. $a + b$
6. $n + 5/n$ or $1 + 5/n$
7. correct
8. $2x^3$
9. correct
10. $5a + 7b$

Daily Warm-Ups: Algebra, NCTM Standards

Answer Key

112. **Mountain bike sales**

	Trek	K2	Schwinn
Los Angeles	100	55	35
Boston	45	40	40
Chicago	30	60	25

Boston sales

	hybrid
Trek	30 45
K2	35 40
Schwinn	50 40

Combined sales all cities

	hybrid	mountain
Trek	135	175
K2	125	155
Schwinn	270	100

Students used addition and subtraction of matrices to construct the matrices.

113.
1. $x = 3.49$
2. $x = 1{,}194$
3. $9\log 2 + 3\log 5$
4. $8\log x$
5. $x = -4$

114.
1. $y = \frac{2}{x}$
2. $y = \frac{1}{2x}$
3. $y = 2x^3$
4. $y = 1$
5. $y = x^2$
6. $y = 3^5 = 243$
7. $y = 5^4 = 625$

115. $45 = 22 + 23 = 14 + 15 + 16 = 7 + 8 + 9 + 10 + 11$; 64 has no consecutive integer sums; $80 = 14 + 15 + 16 + 17 + 18$.

116.
1. growth; R = 3; 300% increase
2. growth; R = 1.1; 110% increase
3. decay; R = .6; 60% decrease
4. decay; R = .2; 20% decrease
5. decay; R = .38; 38% decrease

117. Students might discuss how the equation $x^2 = -9$ cannot be solved in the set of real numbers because the square of any number is a nonnegative number. Students might also talk about the graph of the equation $y = x^2 + 9$ and show that it does not intersect the x-axis, so there are no "zeros" or roots for $y = 0$.

Daily Warm-Ups: Algebra, NCTM Standards

118.
1. speed = 1.25 kilometers per minute
2. $d = 1.25t$
3.

Time	20	40	60	80	100
Distance	24.5	50	75	100	125

4. 24.5 kilometers
5.

Speed Traveled

119. The goal is to show that the following sentence is true for n, where n is any real number:
$$n^2 + (n + 1)^2 + [n(n + 1)]^2 = [n(n + 1) + 1]^2$$
Starting on the left side of the equation:
$n^2 + (n + 1)^2 + [n(n + 1)]^2$
$= n^2 + [n^2 + 2n + 1] + [n^2 + n]^2$
$= 2n^2 + 2n + 1 + [n^4 + 2n^3 + n^2]$
$= n^4 + 2n^3 + 3n^2 + 2n + 1$
On the right side of the equation:
$[n(n + 1) + 1]^2$
$= [n^2 + n + 1]^2$
$= n^4 + 2n^3 + 3n^2 + 2n + 1$

120. Both arithmetic and geometric sequences are represented.
1. $t_1 = 1, t_n = 3(t_{n-1})$
2. $t_1 = 2, t_n = t_{n-1} + 7$
3. $t_1 = 36, t_n = \frac{t_{n-1}}{3}$
4. $t_1 = -12, t_n = t_{n-1} + 5$

Daily Warm-Ups: Algebra, NCTM Standards

Answer Key

121.
1. $t_1 = 1$, $t_n = t_{n-1} + 4$, $n \geq 2$; tenth term = 37; arithmetic sequence
2. $t_1 = 1$, $t_2 = 1$, $t_n = t_{n-1} + t_{n-2}$, $n \geq 3$; tenth term = 55; neither (Fibonacci)
3. $t_1 = 10$, $t_n = \frac{1}{10}t_{n-1}$, $n \geq 2$; seventh term = .0001; geometric sequence
4. $t_1 = 0$, $t_n = t_{n-1} + \frac{1}{6}$, $n \geq 2$; eleventh term = $\frac{5}{3}$; arithmetic sequence
5. $t_1 = 2$, $t_n = t_{n-1}(2)$, $n \geq 2$; fifteenth term = $-32,768$

122.
1. $-18, -12, -6, 0, 6, 12, 18, \ldots$
2. $.5, 1.5, 4.5, 13.5, 40.5, 121.5, 364.5, \ldots$
3. $37.6, 33.4, 29.2, 25, 20.8, 16.6, 12.4, \ldots$
4. $6, -3, +1.5, -0.75, +0.375, -0.1875, +0.09375, \ldots$

123. $y = \pm \frac{1}{2}\sqrt{x+4}$

t	x	y
-5	21	-2.5
-4	12	-2
-3	5	-1.5
-2	0	-1
-1	-3	-0.5
0	-4	0
1	-3	0.5
2	0	1
3	5	1.5
4	12	2
5	21	2.5

124. Original equation: $y = f(x) = x^2 + 8x + 10$
Roots: $x = -4 \pm \sqrt{6}$; x-intercepts are $(-4-\sqrt{6}, 0)(-4+\sqrt{6}, 0)$.

125. Parametric equations show the relationship between two variables, usually x and y, and a third variable, usually t. They are often used to model situations involving two different moving objects (x and y) as they change over time (t).

Daily Warm-Ups: Algebra, NCTM Standards

126. 1. The graph is an ellipse with horizontal major axis 10 units long and vertical minor axis 4 units long.
    2. $x = 5\cos(t)$ and $y = 2\sin(t)$
    3. The new equation would be
    $$\left(\frac{x+1}{5}\right)^2 + \left(\frac{y-2}{2}\right)^2 = 1.$$
    The parametric equations would be $x = 5\cos(t) - 1$ and $y = 2\sin(t) + 2$.

Part 5: Understanding Patterns, Relations, and Functions

127. 1. 40 Popsicle sticks
    2. explicit equation: $t_n = 3(n-1) + 4$; recursive equation: $t_1 = 4$, $t_n = t_{n-1} + 3$
    3. Polly's pattern could not contain *exactly* 71 Popsicle sticks. The closest thing would be 70 Popsicle sticks—a chain of 23 parallelograms.

128. The fifth figure will have 15 dots. The sixth figure will have 21 dots. The equation is $T = \frac{n(n+1)}{2}$, where T is the value of the triangular number and n is the number of the term.

129. 1. This is a representation of square numbers.
    2. 100 blocks
    3. 10,000 blocks
    4. $B = n^2$

130. You might want to encourage students to think about square numbers and triangular numbers.
    $$P_n = S_n + T_n = n^2 + \frac{n(n-1)}{2} = \frac{n(3n-1)}{2}.$$

131. 1. 40 pages
    2. 20 and 21; 41
    3. One possibility: N sheets of paper result in $4N$ pages. The sum of the page numbers on one side of 1 sheet is $4N + 1$.
    4. 820
    5. $2N$ represents the number of sides of each sheet. Thus, the sum of the pages $= 2N(4N + 1)$.

132. One way to think about this situation is to imagine the stacked boxes are a staircase that is half of a rectangle, where n represents the base and $n + 1$ represents the height of the rectangle. The value $2B$ then would be equal to $n(n + 1)$. In other words, twice the number of boxes must equal the product of two consecutive

Daily Warm-Ups: Algebra, NCTM Standards

Answer Key

integers. Solving for the equation $2B = n(n + 1)$ where $B = 45$ gives a quadratic where $n = -10$ or $n = 9$. Since Freya can't use a negative number of boxes, she should use 9 boxes on the bottom row.

133. Paolo could make 29 pieces with 7 cuts.

Number of cuts	0	1	2	3	4	5	6	7	8	9	n
Number of pieces	1	2	4	7	11	16	22	29	37	46	$\frac{n(n+1)}{2}+1$

134. 1. 625; no middle number in row 50; n^2
 2. 6; 8; 78; $2(n-1)$
 3. 29; 41; $n^2 + n - 1$
 4. n^3
 5. $\left[\frac{n(n+1)}{2}\right]^2$

135. $D = 60t$. Letters will vary.

136. Answers will vary. Sample answers: $4(S + 1)$; $2S + 2(S + 2)$; $(S + 2)^2 - S^2$; $4(S + 2) - 4$; $S + S + S + S + 4$

137.
U-Say	3	0	−4	1	2	5
I-Say	11	2	−10	5	8	17

Abby's rule: $U = 3 \cdot I + 2$

138. 1. $x - y = 12$, $xy = 45$, thus $(12 + y)y = 45$, $y^2 + 12y - 45 = 0$, $(y + 15)(y - 3) = 45$, thus the values are 3 and 15, or −3 and −15

 2.

Daily Warm-Ups: Algebra, NCTM Standards

Answer Key

3. Students might choose values of x and y whose difference is 12 and test for the products. This strategy might not lead them to the negative values that satisfy this situation.

x	y	Product
13	1	13
14	2	28
15	3	45

4–5. Answers will vary.

139. 1. linear regression $(a + bx)$
     regEQ$(x) = 8.1875 + .458333x$
     2–3. Answers will vary.

140. 1.

2.

141. Graph A looks like it might be linear. Graph B looks like a curve that opens upward; as x increases, y decreases. Graph C does not seem to have a best-fit line or curve.

142. 1. quadratic
     2. linear
     3. quadratic
     4. exponential
     5. quadratic

143. 1. arithmetic: $t_0 = 30$, $t_n = t_{n-1} + 5$, for $n \geq 1$
     2. geometric: $t_0 = 80$, $t_n = at_{n-1} + 40$, for $n \geq 1$

144. Answers will vary.

145. Zeros: 3, -1; vertex: $(1, -4)$; intercepts: $(0, -3)$, $(3, 0)$, $(-1, 0)$. Other answers will vary.

146. Points will vary.
     1. $y = |x|$
     2. $y = x^2$
     3. $y = \dfrac{1}{x}$
     4. $y = x^3$

Daily Warm-Ups: Algebra, NCTM Standards

Answer Key

147. 1. a. $y = 2x + 7$; As x increases, y increases by a constant amount. First differences are constant.
     b. missing values: (4, 15) (5, 17)
     c. The relationship is linear.
  2. a. $y = -x^2 + 11x$; As x increases, y increases by a decreasing amount. Second differences are constant.
     b. missing values: (4, 28) (5, 30)
     c. The relationship is quadratic.

148. 1. a. $y = 3^x$; As x increases, y increases by a factor of 3. There is a constant growth factor.
     b. missing values: (4, 81) (5, 243)
     c. The relationship is exponential.
  2. a. $y = -x^2 + 4x$; As x increases, y increases by a decreasing amount. Second differences are constant.
     b. missing values: (4, 0) (5, −5)
     c. The relationship is quadratic.

149. 1. a. $y = x^2 - 5x$; As x increases, y decreases by a decreasing amount. Second differences are constant.
     b. missing values: (−1, 6) (5, 0)
     c. The relationship is quadratic.
  2. a. $xy = 1/2$; As x increases, y is $1/2x$.
     b. missing values: (4, 1/8) (5, 1/10)
     c. The relationship is inverse.

150. Answers will vary. Sample answer: Not all quadratic functions have x-intercepts, but they do all have a y-intercept. Quadratic functions can have 2, 1, or 0 x-intercepts depending on the location of the vertex and the direction of the opening. One example: If the quadratic function has a vertex below the x-axis and opens down, then it has no x-intercepts.

151. 1. b; the degree of the function is even, meaning it exhibits the same end behaviors, and the coefficient 8 causes the graph to shrink compared to its parent graph of $y = x^2$
  2. a; the degree of the function is odd, meaning it exhibits opposite end behavior, and the coefficient of .1 causes the graph to stretch compared to its parent graph of $y = x^3$
  3. d; the degree of the function is even, meaning it exhibits the same end behavior
  4. c; the degree of the function is odd, meaning it exhibits opposite end behavior

Daily Warm-Ups: Algebra, NCTM Standards

Students may justify their answers in a variety of ways. They may indicate points on the lines that fit the equations. They may talk about the steepness of the lines of the rates of change of the graphs.

152.
1. opens up; vertex $(0, -5)$; axis of symmetry $x = 0$
2. opens up; vertex $(1, 2)$; axis of symmetry $x = 1$
3. opens up; vertex $(-3, -1)$; axis of symmetry $x = -3$

153.
1. Both are parabolic functions with vertex at $(0, 0)$.

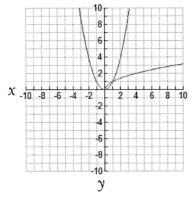

2. No, the graph of $y = \pm\sqrt{x}$ would not pass the vertical-line test and would not be a function.

154. Answers will vary. Sample answer: If $b = 0$ and $c = 0$, the graph of $y = x^2$ will stretch or shrink and will be symmetric to the y-axis. If $a < 0$, the graph will open downward. If $a > 0$, the graph will open upward. If $a = 1$ and $b = 0$, the value c represents the y-intercept, and the graph $y = x^2$ will be transposed along the y-axis.

155.
1. The graph will slide up 3 units.
2. There will be no change in the graph.
3. The graph will slide right 3 units.
4. The graph will shrink by a factor of 2 and will slide down 1 unit.

156.
1. The graph has been slid left 4 units; $y = (x + 1)^2$
2. The graph has been slid left 2 units and flipped or reflected over the x-axis; $y = -(x - 1)^2$
3. The graph has been reflected over the x-axis; $y = -(x - 3)^2$
4. The graph has been slid 6 units left and reflected over the x-axis (order can be different); $y = -(x + 3)^2$

Daily Warm-Ups: Algebra, NCTM Standards

Answer Key

157. 1. Penny's graph is a little off. The x-intercepts for this graph should be (6, 0) and (−2, 0). The vertex is at (2, −16) rather than (−2, −16).
     2. vertex form: $y = (x − 2)^2 − 16$; polynomial form: $y = x^2 − 4x − 12$

158. Equations 1 and 3 are parabolas with minimum points. In each, x is squared and the coefficient of x is positive. Equations 2 and 4 are also parabolas, but with maximum points; e is linear.

159. 1. −1/2 and 3
     2. $x = −1$
     3. $\dfrac{-2 \pm \sqrt{-8}}{2}$, which means there is no intersection with the x-axis
     4. $\dfrac{-3 \pm \sqrt{17}}{4}$

     For all quadratic functions, the roots or zeros represent the intersection of the graph with the x-axis.

160. These are all parabolas or quadratic functions. Their rates of change switch from decreasing to increasing (or increasing to decreasing) at their vertex, which represents their maximum or minimum point. The coordinates of the vertices for these functions are (−3.5, −8.25) and (3.5, 16.25). The axis of symmetry passes through these vertices such that $x = -b/2a$. The y-intercepts are (0, 4) for the first two equations and (0, c) for the third equation.

161. 1. no x-intercepts
     2. two x-intercepts
     3. one x-intercept
     4. two x-intercepts

     Students can determine answers using discriminants.

Daily Warm-Ups: Algebra, NCTM Standards

Answer Key

162.
1. $x \neq 0, y \neq 7$, vertical asymptote: $x = 0$, horizontal asymptote: $y = 7$
2. $x \neq 5, y \neq 3$, vertical asymptote: $x = 5$, horizontal asymptote: $y = 3$
3. $x \neq 0, y \neq 2$, vertical asymptote: $x = 0$, horizontal asymptote: $y = 2$

163. Answers will vary. The graphs of a and c lie completely in the first and third quadrants, and the graphs of b and d lie completely in the first and second quadrants. All approach zero as x approaches infinity. Asymptotes are $x = 0$ and $y = 0$ for all graphs. All are functions.

a. $r(x) = \dfrac{3}{x}$

b. $r(x) = \dfrac{3}{x^2}$

c. $r(x) = \dfrac{3}{x^3}$

d. $r(x) = \dfrac{3}{x^4}$

Daily Warm-Ups: Algebra, NCTM Standards

Answer Key

164. 1. As x gets very large positively, the functions increase rapidly. The y-intercept is $a + c$.
     2. When a is negative, the graphs will reflect over the x-axis. Changing b to $1/b$ reflects the graph over the y-axis.

165. 1. y-intercept: $(0, 16)$; x-intercepts: $(-4, 0)$, $(4.3, 0)$. Other answers will vary.
     2. y is increasing over the interval $(-5, 0)$.
     3. No, the rate of change is not constant. The change over $(0, 2)$ is -1, over $(2, 4)$ is $-7/2$, and over $(4, 5)$ is -7.

166. Rita is confused about the definition of a function. While it is true that the line $y = 5$ intersects the graph of each equation in two places, this indicates that for any y value there may be two values of x. Both equations are functions because a vertical line cannot intersect these graphs in any more than one point. Thus for any value of x, there is one and only one value of y.

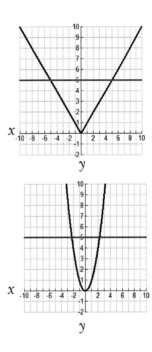

Daily Warm-Ups: Algebra, NCTM Standards

167. The functions represent exponential functions. For equations of the form $h(x) = a^x$ and $a > 1$, $h(x)$ is increasing. For $0 < a < 1$, $h(x)$ is decreasing. An asymptote exists at $y = 0$. The graphs have a y-intercept at $(0, 1)$. There are no maximum or minimum values, and there are no symmetry features.

168. The function is called a *step function* because that's what it looks like on the graph. It is also known as the greatest integer function because the value of x is equal to the smallest integer not greater than x in the sentence $y = [x]$.

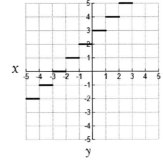

169. Answers will vary. A discussion of rational functions is likely. Some equations that would create a hole or an asymptote at $x = 2$ are

$$y = \frac{(x-2)^2}{x-2}, y = \frac{1}{x-2}, y = \frac{1}{(x-2)^2}, \text{ and } y = \frac{x-2}{x-2}.$$

Daily Warm-Ups: Algebra, NCTM Standards

Answer Key

170.
1. For very large or very small values of x, the function approximates the graph of $g(x) = 2x + 17$. $f(x)$ can be rewritten as $g(x) + h(x)$. Carol's viewing window looks like $g(x)$.
2. Each friend has a different window for the graph of the function.
3. The function is undefined at $x = 2$, and there is an asymptote for the graph at that vertical line.

171. Students should indicate that they understand about interchanging the independent and dependent variables to create an inverse and that $f^{-1}(f(x)) = x$. They might also discuss that inverses are reflections over the line $y = x$.
1. $f^{-1}(x) = \frac{5}{2}x - 3$
2. $g^{-1}(x) = (x + 6)^{3/2} - 2$
3. $h^{-1}(x) = \log_2(6 - x)$

172.
1.

x	$f(x)$	$g(x)$	$f(g(x))$	$g(f(x))$
–1	2	4	17	1
2	5	1	2	16
3	10	4	17	81

2. $f(g(x)) = ((x – 1)^2)^2 + 1$; $g(f(x)) = ((x^2 + 1) – 1)^2$

173.
1. –4
2. 2
3. $x^2 – 3$
4. –16
5. 14
6. $–3a^2 + 11$
7. $(–3a + 4)^2$
8. $(x – 1)^4 – 2$

174.
1. $F(–1) = 0$
2. $F(5) = 0$
3. $F(x) = 3$ for $1 \leq x \leq 3$ and $x = 5.75$
4. $F(x) > 4$ for $6 \leq x \leq 8$
5. domain: $–1 \leq x \leq 7$; range: $0 \leq y \leq 7$

Daily Warm-Ups: Algebra, NCTM Standards

Answer Key

175. Tables *a* and *c* could be functions. Tables *b* and *d* could not. In table *b*, x = 3 is paired with two different values of y. In table *d*, x = a is paired with two different values of y. This contradicts the definition that for any *x* there will be one and only one *y*.

176. In Graph 1, $0 < a < 1$ and $C > 0$. In Graph 2, $b > 1$ and $D > 0$.

177.

x	y
−2	1,600
−1	320
0	64
1	12.8
2	2.56
3	.512
4	.1024

A graph of the function would look as follows. This represents an exponential decay model:

Equation: $y = 64(1 - 0.8)^x$

Daily Warm-Ups: Algebra, NCTM Standards

Answer Key

178. Predictions will vary.

original graph:

graph a:

graph b:

graph c:

179. 1. $y = \cos(x)$

2. $y = \cos(x)$

3. $y = \cos(x)$
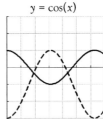

180. Answers will vary. Possible solutions: The dashed-line graph looks like it might be shifted 1 unit to the right; the function $f_2(x) = (x - 1)^4 + (x - 1)^3 - 3(x - 1)^2$ would create this situation. The dotted-line graph is shifted right, but it's also moved 5 units up, so the function could be $f_3(x) = (x - 1)^4 + (x - 1)^3 - 3(x - 1)^2 + 5$.

Daily Warm-Ups: Algebra, NCTM Standards